CONCRETE MANUAL
Study Companion

Updated to
2006 IBC®
AND
ACI 318-05

ICC
INTERNATIONAL
CODE COUNCIL®

2006 Concrete Manual Study Companion

Based on the 2006 IBC and ACI 318-05

ISBN-13: 978-1-58001-502-8
ISBN-10: 1-58001-502-6

Cover Design: Mary Bridges
Publications Manager: Mary Lou Luif
Project Editor: Roger Mensink
Typesetting: Lis Valdemarsen

Copyright © 2006, *International Code Council*

First Printing: November 2006

Printed in the United States of America

Introduction

This study companion is intended to provide practical learning assignments for independent study of the *Concrete Manual*. The independent study format provides a method for the student to complete the study program in an unlimited amount of time. Proceeding through the workbook, the student can measure his or her level of knowledge by using the quizzes in each study session.

All study sessions contain specific learning objectives, a list of statements and questions summarizing the key points for study, and quizzes designed to assess the student's retention of technical knowledge. Therefore, before beginning the quizzes, students should thoroughly review the corresponding chapter of the *Concrete Manual* concerning the learning objectives and key points.

The quizzes are designed to encourage the student to develop the habit of carefully reading the text for a clear understanding of the subject material. The questions are not intended to be tricky or misleading. The following three formats are used to vary the method of evaluation:

1. Multiple Choice—Each statement is followed by a unique group of possible responses from which to choose.

2. True/False—Each statement is either true or false.

3. Completion—Each statement must be correctly completed by inserting the proper *Concrete Manual* text.

The study companion is structured so that every question is followed by the opportunity to record the student's response and the corresponding text reference. The correct responses are indicated at the back of the workbook in the answer key so that the student can assess his or her knowledge immediately.

Many new questions have been added to reflect the 2006 *International Building Code*® and the 2005 edition of ACI Standard *Building Code Requirements for Structural Concrete* (ACI 318-05).

ACKNOWLEDGMENTS

The International Code Council® would like to extend its appreciation to Donald M. Hunsicker for his preparation, under special contract to ICC®, of the original text materials for this study guide. Mr. Hunsicker's development of this unique study aide provides an excellent resource to those individuals involved in the inspection of concrete.

Mr. Hunsicker is currently assistant building official with the City of Visalia, California. He has been active in the construction field for over 25 years, with the last 14 years being dedicated to the field of building inspection. Mr. Hunsicker holds degrees in Building Inspection Technology and Vocational Education. His writing credits include two other inspection-related workbooks published by ICC.

TABLE OF CONTENTS

CHAPTER 1
FUNDAMENTALS OF CONCRETE

Objectives: To outline a brief history of cement and concrete, describe the hydration process, identify the characteristics of concrete, introduce the role of admixtures and the water-cement ratio, define "good durable concrete" and the causes of distress and failure, and briefly discuss the five fundamentals of concrete.

Lesson Notes: Special attention should be given to the nine properties of good durable concrete (they will be discussed in detail in subsequent chapters) and their relationship to the five fundamentals of concrete construction.

Key Points:

- What was the first type of masonry?
- Name the different types of cementing materials developed prior to the Middle Ages.
- From where does the term "pozzolan" originate?
- What was John Smeaton's contribution to the development of cement?
- Who first developed portland cement?
- Where was the first portland cement in the United States produced and by whom?
- What led to the large-scale production of cement?
- What is the first basic law of concrete technology, and who did the research and observations that established it?
- Approximately when were the following introduced? High-frequency vibrators, low-heat cement, precast concrete, admixtures and low-alkali cement.
- Identify the important developments in concrete over the past 25 years.
- Describe the hydration process.
- For how long will the hydration process continue?
- What affects the rate of hydration?
- What is generated during hydration?
- What is concrete?
- What is mortar? Grout?
- How do concrete and steel compliment each other?
- What are the characteristics of fresh concrete?
- Define green concrete.
- Describe the water-cement ratio law. How absolute is this law?
- What other factors can contribute to concrete strength and durability?
- In what ways do admixtures modify concrete's properties?
- Define the properties of "good durable concrete."
- Name three general reasons for the distress or failure of concrete.
- Are distress or failure usually caused by only one factor?
- What facts should be considered when investigating a failure?
- Name the five fundamentals of concrete construction.
- What is the most probable cause of distress in concrete?
- Why is site investigation important?
- How might a poor structure design manifest itself?
- In what ways can materials and mixes best be judged and selected?
- How important is workmanship?
- What does the term "workmanship" mean?
- How does maintenance affect a structure?
- How should the five fundamentals be applied?

CHAPTER 1–QUIZZES

I Multiple Choice

1.   Who developed the first portland cement by burning limestone and clay at high temperatures?
     a. Romans
     b. Aspdin
     c. Eddystone
     d. Smeaton   *1824*
     e. Greeks
     Response ___b___   Reference _pg 3 1a paragraph_

2.   Which of the following is not one of the five fundamentals of good durable concrete?
     a. material selection
     b. proper structure design
     c. reasonable cost
     d. site investigation
     e. workmanship
     Response ___c___   Reference _pg 7_

3.   Fresh concrete is ___b___.
     a. green
     b. plastic
     c. newly placed
     d. self-supporting
     e. none of the above
     Response _____   Reference _____

4.   Hydration produces ___a___.
     a. heat
     b. water
     c. drying
     d. cooling
     e. shrinkage
     Response _____   Reference _____

5.   The first law of concrete to be researched and observed is the ___e___.
     a. hydration rate
     b. admixture reaction
     c. drying time/strength
     d. volume stability
     e. water-cement ratio
     Response _____   Reference _____

II True/False

6. Admixtures provide a way to achieve certain properties in concrete.
    T _✓_ F _____ Reference _____

7. Good workmanship includes proper material selection.
    T _✗_ F _✓_ Reference _____

~ 8. Distress is only a maintenance concern.
    T _____ F _✓_ Reference _____

~ 9. Investigation of materials for the Hoover Dam resulted in development of low-heat concrete.
    T _✓_ F _____ Reference _____

— 10. Portland cement is composed only of lime and clay.
    T _✓_ F _✗_ Reference _____

III Completion

11. Concrete that has low strength, high moisture content and is only a few hours or days old is referred to as ___GREEN___ concrete.
    Reference _____

12. The property of concrete that resists attack by weather or substances is ___Durability___. Reference _____

13. The forces of weather can be destructive to concrete by ___Freeze___ and ___THaw___ that produce cracks, followed by entrance of ___WATER___ into the cracks. Reference _____

—14. Burnt ___GYPSum___ was first developed in early Egypt.
    Reference _____

— 15. The ___Rotary Kiln___ led to large scale production of cement worldwide.
    Reference _____

CHAPTER 2
THE FRESH CONCRETE

Objectives: To obtain an understanding of the significance of workability, how it is measured, the factors affecting it and the concurrent properties of segregation, bleeding, unit weight and air content.

Lesson Notes: "Consistency," "cohesiveness" and "plasticity" are terms that are interrelated but describe different aspects of concrete's workability. Consistency is a measure of wetness or fluidity. Cohesiveness indicates whether concrete is harsh (low adhesion), sticky (high adhesion) or plastic (good adhesion and not easily segregated). Plasticity is the quality of fresh concrete that allows concrete to be molded or formed into a final configuration without segretation when properly handled.

Key Points:

- Define "workability" and "plasticity" and describe the differences between them.
- On what properties does workability depend?
- Why does unworkable concrete increase cost?
- What might occur if minimum effort is put into consolidation of a high-slump concrete?
- Name the two optimum conditions for concrete in normal structures.
- What three terms are used to describe the workable aspects of concrete?
- Define "consistency."
- How is consistency measured?
- How can fluidity be increased?
- What effect does temperature have on slump?
- What is meant by the term "cohesiveness"?
- What does a harsh concrete lack?
- Name the causes of harshness.
- How can harshness be remedied?
- Where might a harsh concrete mix be desirable?
- What is a common occurrence in a sticky mix?
- Do sticky mixes segregate easily? Explain.
- What test is most practical for field work?
- What test measures consistency?
- Identify the factors that can affect workability.
- What type of cement can be used to increase workability?
- What is meant by the term "false set"?
- In what manner should it be treated?
- What is meant by the term "flash set"?
- Identify the cause.
- Which aggregate is more desirable—natural gravel or crushed rock? Why?
- What type of aggregate produces poor workability?
- How does aggregate affect workability?
- How might admixtures affect workability?
- What is the range of particle sizes in concrete materials?
- Define "segregation."
- In hardened concrete, what can be the result of segregation?
- Which type of mixes tend to segregate?
- During which phases is it critical to control segregation?
- What is bleeding?
- Where does it occur most frequently?
- For how long might it continue?

- What can influence bleeding?
- Name the detrimental effects of too much bleeding.
- What can be done to limit or stop bleeding?
- What is laitance and of what does it consist?
- How much strength does laitance have?
- How would laitance affect a joint or fill?
- Identify the causes of laitance.
- What measures can reduce laitance?
- Define unit weight and yield
- What are the factors that affect unit weight?
- What does the unit weight indicate?
- When is determining the amount of air present in a mix important?
- Name the three ways to measure air content.
- Describe how each works.
- Which is the most accurate?
- How does air-entrainment affect concrete?
- Besides admixtures, what variables can affect the amount of air in a mix?

CHAPTER 2–QUIZZES

I Multiple Choice

1.   The amount of air in all concrete is ____a____ percent.
         a. between one and two
         b. at least three
         c. as much as five
         d. a maximum of eight
         e. as high as ten
         Response _____ Reference _____

- 2.   The element of workability that indicates whether concrete is plastic, sticky or harsh
         is ____a____.
         a. cohesiveness
         b. consistency
         c. slump
         d. air content
         e. water content
         Response _____ Reference __2.2__ pg 12__

- 3.   Which of the following is not a result of segregation?
         a. rock pockets
         b. laitance
         c. sand streaks
         d. bleeding
         e. scaling
         Response _____d_____ Reference _____

- 4.   A harsh and unworkable concrete can result from ____d____.
         a. finely ground cement.
         b. fine aggregate
         c. adding pozzolans
         d. low cement content
         e. rounded or subrounded aggregates
         Response _____ Reference _____

5.   A sticky concrete mix usually contains a high ____c____ content.
         a. air
         b. aggregate
         c. cement or rock dust
         d. water and pozzolans
         e. pozzolans
         Response _____ Reference _____

II True/False

6. The most important property of fresh concrete is workability.
   T _✓_ F _____ Reference _____

7. The kind of structural element does not determine workability.
   T _____ F _✓_ Reference _____

8. Grading of coarse aggregate is more critical then grading of fine aggregate.
   T _____ F _✓_ Reference _____

9. Entrained air can cause separation.
   T _____ F _✓_ Reference _____

10. A well-graded sand usually procedures a low bleeding rate in concrete.
    T _✓_ F _____ Reference _____

III Completion

11. Air-entrainment can improve workability, lower ___unit wt___ and reduce ___Bleeding___. Reference _____

12. The weight of one cubic foot of concrete is referred to as ___u wt___.
    Reference _____

13. A lack of _____ or _____ will cause segregation in concrete that flows into place.
    Reference _____

14. Harsh, low slump mixes are most common when used for _____, _____ and _____.
    Reference _____

15. ___Bleeding___, especially in flat slabs, is accompanied by a slight settlement of solid particles.
    Reference _____

CHAPTER 3
THE STRENGTH OF CONCRETE

Objectives: To understand the importance of strength, the kinds of strength, how strength is measured and the various factors affecting strength.

Lesson Notes: Concrete is well known for its compressive strength. However, there are many factors that may affect this strength. By examining Table 3.3, you will gain an understanding into the causes and effects of some of these factors.

Key Points:

- How is the quality of concrete judged?
- At what age is concrete tested?
- What is the basis for acceptance or rejection of concrete?
- Who determines the strength of structural concrete?
- Other than strength, what properties of concrete can be significant?
- On what do these properties depend?
- What are the different types of strengths that may be associated with concrete?
- For what type of loading is concrete a good material?
- What is the standard size cylinder for testing compressive strength?
- What types of elements are loaded in flexure?
- What part of a concrete member is the weakest?
- Define the "modulus of rupture."
- What test is a good indicator of tensile strength in concrete?
- Name the stresses that concrete is nearly always subjected to.
- Identify the factors that may affect concrete strength in a structure.
- What are the four basic methods by which concrete can be tested?
- What is a job-molded specimen?
- How does a Swiss hammer work?
- Describe how a Windsor probe tests concrete strength.
- How accurate are these two instruments?
- Describe how sonic testing is accomplished.
- What is one problem with strength tests?
- How does a high water content affect concrete?
- In what ways might cement contribute to low strength concrete, and how can its effect be minimized?
- How do aggregates affect strength?
- When are larger aggregates used?
- When are smaller aggregates used?
- Identify the three relationships between aggregates and concrete strength.
- How are deficiencies in aggregate remedied?
- In what way do clay coatings affect concrete strength?
- What are considered to be the maximum amounts of rock dust or other fine materials acceptable in coarse and fine aggregate?
- How should organic matter be dealt with?
- How does moisture in aggregate affect concrete?
- What is the lowest acceptable specific gravity in aggregates?
- How much organic material in water may cause a serious deficiency in strength?
- What types of chemicals are not acceptable in concrete mixing water?
- Is the volumetric measurement of ingredients good practice? Explain.

- Name the batching errors that may contribute to reduced concrete strength.
- At what range of temperature is strength retarded, unfavorable to early strength or absent?
- What is considered the optimum temperature for placing concrete?
- Describe how freezing affects strength.
- When is rapid strength development advantageous?
- Name the five methods to accelerate strength.
- What type of cement is high-early, and how does it differ from other cements?
- When is calcium chloride acceptable as an admixture? When is it not?
- How might insulating forms contribute to curing?
- Where is high-temperature curing most frequently used?
- What type of materials are produced using high-temperature curing methods?
- Identify the different compounds used in rapid-setting cements.
- How would an overdose of retarder affect concrete strength?
- What are the effects of sugar compounds on strength, and how much does it take to affect strength?
- What occurs when concrete is placed and kept at near freezing?
- At what psi is concrete considered high strength?
- Where might high-strength concrete be used?
- At what age are specimens of 10,000 psi concrete tested?
- How should cement, aggregate, admixtures and mixes be selected?
- How cool should concrete be kept?
- Identify the quality control measures associated with concrete.
- What types of concrete structures are subjected to high temperatures?
- How might tests that determine the effects of heating vary?
- What is considered to be the maximum heat that concrete should be exposed to on a continuing basis when one surface is exposed or when members are light bearing?
- When all surfaces are exposed?
- In what way can fire damage concrete?
- How might such damage be repaired?

CHAPTER 3–QUIZZES

I Multiple Choice

1. Compressive strength tests are usually done when the sample has been aged for ___28___ days.
   a. seven
   b. 14
   c. 21
   (d.) 28
   e. 35
   Response _____ Reference _____

2. Other factors aside, the best range of temperature for placing concrete is between _____ degrees F.
   a. 20 to 60
   (b) 40 to 80
   c. 50 to 90 (50-80) const. so CONTRACTOR
   d. 60 to 90
   e. 40 to 120
   Response _____ Reference _____

3. High-early cement is made by increasing the amount of tricalcium silicate and ____e____.
   a. calcium chloride
   b. hydration
   c. air-entrainment
   d. high-temperature curing
   (e) finer grinding of the cement
   Response _____ Reference _____

4. MSA stands for ___b___.
   a. modified strength admixture
   (b) maximum size aggregate
   c. modulus of shear axial
   d. minimum size aggregate
   Response _____ Reference _____

5. Mineral admixtures used to achieve strengths between 8,000 and 20,000 psi are _____.
   a. pozzolans and chert
   b. chert and ground manganese
   c. caliche and rock dust
   d. fly ash and silica fume
   e. calcium and aluminum silicate
   Response _____ Reference _____

6. A type of aggregate that should be avoided due to its effects on strength is

_____.
a. crushed quartz
b. any of glacial origin that contains organic matter
c. any with a high specific gravity
d. granite
e. all of the above
Response _____ Reference _____

7. Test specimens are valuable in that they give a measure of _____ and other
properties of concrete.
a. specific gravity
b. strength potential
c. density
d. tension resistance
e. all of the above
Response _____ Reference _____

8. A source of batching errors is _____.
a. careless operation
b. allowance for moisture variables in aggregates
c. scales returning to zero between batches
d. placing methods
e. all of the above
Response _____ Reference _____

9. Compressive strength of precast and prestressed concrete is typically specified to
be _____ psi.
a. below 2,000
b. 2,500 to 3,500
c. 3,000 to 4,000
d. 4,000 to 7,000
Response _____ Reference _____

10. The strength of concrete most commonly measured is __a__.
a. compressive strength
b. bending of flexural strength
c. tensile strength
d. none of the above
Response _____ Reference _____

11. As compared to the compressive strength measured by a 6-inch diameter by 12-inch high cylinder, the compressive strength of a 4-inch diameter by 8-inch high cylinder will generally be _____.
    a. significantly lower
    b. slightly lower
    c. about the same
    d. higher
    Response _____ Reference _____

12. The modulus of rupture of concrete is a measure of the _____.
    a. compressive strength
    b. tensile strength
    c. flexural strength
    d. shear strength
    Response _____ Reference _____

13. Tensile strength of concrete can be measured indirectly by a _____.
    a. compressive strength test
    b. flexural strength test
    c. split cylinder test
    d. direct tension test
    Response _____ Reference _____

14. Cores taken from near the top of a column will generally indicate __b__ strength, compared with cores taken from near the bottom of the same column.
    a. higher
    b. lower
    c. about the same
    d. slightly higher
    Response _____ Reference _____

15. Commonly, the major cause of compressive strength test variation is _____.
    a. cement composition variations
    b. water-cement ratio variations
    c. cement temperature variations
    d. mixing speed variations
    Response _____ Reference _____

16. Concrete made with 660 pounds of cement per cubic yard with well-graded aggregates having a maximum size of _____ will have the highest strength.
    a. $^3/_8$ inch
    b. $^3/_4$ inch
    c. $1^1/_2$ inches
    d. 3 inches
    Response _____ Reference _____

17.   To gain strength rapidly during the first few days after casting, which one of the following can be used?
a. high-early strength cement
b. an accelerating admixture
c. curing at high temperature
d. any of the above
Response _____ Reference _____

18.   Concrete will gain strength slowly if __b__.
a. it contains an overdose of a water-reducing admixture
b. it contains an overdose of a retarder
c. it contains an overdose of an accelerator
d. the concrete and air temperature are 80 degrees F
Response _____ Reference _____

19.   Concrete heated to 800 degrees F for a long period of time and then cooled will have a permanent strength reduction of _____ percent.
a. less than ten
b. 10 to 40
c. about 50
d. 50 to 95
Response _____ Reference _____

20.   According to the *International Building Code*, the specified compressive strength for structural concrete can not be less then_____ psi.
a. 1500
b. 2000
c. 2500
d. 3000
e. none of the above
Response _____ Reference _____

II True/False

21.   There is no field test for direct determination of tension under axial loading.
T _____ F _____ Reference _____

22.   When concrete has to be cored to verify strength, damage to reinforcement is not of concern when the psi is below 3,000.
T _____ F __✓__ Reference _____

23.   A common accelerator admixture that is added to the batch in solution is calcium chloride.
T __✓__ F _____ Reference _____

24.   In general, load-bearing concrete members exposed to continuous heat in excess of 500 degrees F should be avoided.
T __✓__ F __✓__ Reference _____

25. Watertightness is important in nearly all hydraulic structures.
T ✓ F _____ Reference _____

26. Irregularly-shaped natural gravel or cube-shaped crushed rock with a rough and slightly porous surface will give the best bond with the cement paste.
T ✓ F _____ Reference _____

27. Concrete that will be continuously exposed to temperatures greater than 150 degrees F should be laboratory tested to determine if the expected temperature will be detrimental.
T _____ F ✓ Reference _____

28. Concrete made and cured at 50 degrees F will have lower strength at three days but higher strength at 28 days than concrete made and cured at 90 degrees F.
T ✓ F _____ Reference _____

29. Concrete strengths in the range of 6,000 to 10,000 psi at 56 days require new technology.
T _____ F ✓ Reference _____

III Completion

30. If concrete is placed and kept at a near-freezing temperature, the hydration process and strength gain will be _slowed_.
Reference _____

31. The _____ of _____ is a measure of flexural strength and can be determined by bending a small beam by applying a concentrated load at each of the _____ points. This beam is usually _____ by _____ inches in cross section.
Reference _____

32. Five basic methods of achieving accelerated strength are _____ cement, _____, retention of the _____, _____ curing and _____.
Reference _____

33. Aggregates that may be deficient in strength are usually a specific gravity less than _____ or have an absorption exceeding _____ percent.
Reference _____

34. Two nondestructive instruments for checking the strength of concrete are a _Swiss Hammer_ and a _WINDSOR probe_.
Reference _____

CHAPTER 4
THE DURABILITY OF CONCRETE

Objectives: To understand the property of concrete known as durability and the agencies of destruction that affect durability. Also considered are the effects of a marine environment and of hydraulic structures on concrete's durability, as well as the typical problems associated with slabs on ground and prevention of deterioration.

Lesson Notes: When concrete is found to lack durability, by far the most common cause is inferior workmanship; specifically, the use of too much mixing water. A high water content can lead to segregation, laitance, rock pockets, cracking, weak permeable layers and porous concrete. Emphasis should be placed on using only the amount of water specified for the mix.

Key Points:

- Define "durability."
- To what properties of concrete is durability closely related?
- What are the six factors that affect durability?
- Name the three methods of measuring durability.
- What are the signs of accelerated weathering?
- How do abnormal expansion and contraction affect concrete?
- What might the causes be?
- What factors can contribute to accelerated weathering?
- Identify the four general categories of destructive agents.
- What are the necessary steps to protect concrete from these agents?
- Name the conditions that are harmful to concrete.
- What does petrographic examination reveal?
- What substances found in aggregates contaminate or weaken concrete?
- How might selecting the right cement affect durability?
- How do water and the substances in water affect durability?
- What is the most frequent cause of deficiencies in the concrete itself?
- How important is workmanship to durable concrete?
- In what way might mix proportions affect durability?
- What is contained in the ACI Committee 515 report?
- List the substances that attack concrete?
- In what way do ammonium salts, animal wastes and sea water deteriorate concrete?
- Identify a type of structure or situation where each item listed under exposure condition or material in Table 4.1 might occur.
- How does water affect the elements of chemical and mechanical attack?
- How do acids affect concrete?
- What are some of the sources of acids?
- Does oil affect concrete? Explain.
- What effects might fuel spillage from jet aircraft have on concrete?
- Why is calcium chloride an agent of deterioration?
- Which de-icing agents are best and worst for use on concrete?
- Identify the deterioration cause and solution for each of the following: coolers and freezers containing brine solution, food-processing plants, tanning and fermentation plants and farm silos.
- Explain how corrosion of metals affects concrete.
- When combined with concrete, what effect does water have on aluminum, lead, copper, zinc and steel?
- In simple terms, explain what an electrochemical couple is.

- Where is concrete most frequently subjected to high temperatures?
- What are the effects of high temperatures on concrete's durability?
- What measures can increase resistance to spalling at high temperatures?
- How might concrete's durability be affected in chimneys?
- What can be done to remedy these effects?
- In what secondary ways might the durability of concrete be affected?
- What might be early indications of structural damage?
- What are the chief causes of structural damage?
- How are salts formed, and what may increase their concentrations?
- To what is sulfate resistance related?
- What are other factors that affect sulfate resistance?
- Name the types of aggregate that can be alkali-silica reactive.
- Describe the effect of this reaction on concrete.
- If a reactive aggregate must be used, what steps can be taken to minimize this reaction?
- What is the effect of freezing on fresh concrete? Is there a remedy?
- How does frost damage concrete?
- How can frost resistance be enhanced?
- What type of systems are usually subjected to waves and currents?
- Name the three wave types considered when designing structures.
- How does each wave type affect concrete?
- What other considerations need to be addressed?
- What is the effect of tides on structures?
- In what ways do physical and chemical damage occur?
- When should protection of waterfront structures begin?
- How do chamfers and fillets improve resistance?
- What materials enhance resistance?
- In what ways does good workmanship help concrete resist attack?
- What aspects are important in hydraulic structures?
- How are the causes of hydraulic back pressure remedied?
- How does cavitation affect conduits carrying water?
- How is cavitation prevented?
- Define erosion as it relates to concrete.
- How does erosion affect concrete? What are the main causes of slab defects?
- Define "scaling," "spalling," "subsidence," "pumping" and "blowups."
- What are the causes of each?
- Why is air-entrainment of concrete important?
- How is air-entrainment accomplished?
- What two points must be remembered about entrained air with respect to durability?

CHAPTER 4–QUIZZES

I Multiple Choice

1. Cavitation can be caused by __e__ .
   a. surface depressions
   b. surface projections
   c. sharp bends
   d. sudden changes in cross section
   e. all of the above
   Response _____ Reference _____

2. Concrete continually exposed to high temperature is affected primarily by _____ .
   a. frequent spalling
   b. accelerated hardening
   c. a reduction of strength
   d. exhaust gases
   e. high and low temperature extremes
   Response _____ Reference _____

3. Concrete that expands and contracts abnormally may be caused by __d__ .
   a. unsound aggregates
   b. temperature changes
   c. reaction between aggregates and cement
   d. all of the above
   e. none of the above
   Response _____ Reference _____

4. Freezing of concrete in the plastic state will reduce durability, weather resistance and strength by as much as __b__ .
   a. one-fourth
   b. one-half
   c. three-fourths
   d. one-third
   e. two-thirds
   Response _____ Reference _____

5. Poor durability in concrete is rarely caused by __b__ .
   a. water
   b. cement
   c. aggregate
   d. workmanship
   e. mix proportions
   Response _____ Reference _____

6. Which of the following is considered a reactive aggregate?
   a. feldspar
   b. quartz
   c. chert
   d. granite
   e. silica
   Response _____ Reference _____

7. Of the following de-icing agents, which one is not recommended?
   a. calcium chloride
   b. urea
   c. sodium chloride
   d. ammonium sulfate
   e. all of the above
   Response _____ Reference _____

II True/False

8. Pavements of concrete placed in the late fall can be exposed to de-icing salts during the first winter of exposure, provided adequate curing is accomplished.
   T _____ F _____ Reference _____

9. Aluminum is attacked by caustic alkalies when exposed to moist concrete.
   T _____ F _____ Reference _____

10. When a slab is placed directly on a fine-grained, plastic, impervious soil, the presence of moisture may create a condition known as pumping.
    T __✓__ F _____ Reference _____

11. Normal weathering may cause a slight roughening of the surface or rounding of the edges, but is not harmful to good durable concrete.
    T __✓__ F __✓__ Reference _____

12. One strength of concrete is its ability to strongly resist acids.
    T _____ F __✓__ Reference _____

13. Entrained air does not improve the durability and other characteristics of concrete exposed to weather in severe climates.
    T _____ F __✓__ Reference _____

14. Streams may not be a good source of water for concrete if the water contains sulfates, tannic acid, organic materials or sugar.
    T __✓__ F _____ Reference _____

III Completion

15. _fillets_ and _champers_ improve the appearance of a structure, and sharp arris, which is subject to spalling and chipping from moving objects, is avoided.
Reference _____

16. The three types of waves that concrete structures should be designed for are _____, _____ and _____.
Reference _____

17. Movement of paving slabs or blocks on the face of an embankment of reservoirs, sea walls or dams may be caused by _____ back pressure upon sudden _____ of the water level.
Reference _____

18. When dealing with potential or actual attack by chemical elements, either proper attention to produce _____ concrete or some sort of _____ should be provided to separate the concrete from the aggressive materials.
Reference _____

19. The six factors that affect the durability of concrete are the characteristics of the _____, _____, _____, _____ on the structure, structural _____ and construction _____.
Reference _____

20. _____ salts are destructive to concrete because, in the alkaline environment of concrete, they release _____ gas and _____ ions that must be placed by dissolving calcium from the concrete, resulting in a leaching action similar to an _____ attack.
Reference _____

CHAPTER 5
VOLUME CHANGES AND OTHER PROPERTIES

Objectives: To understand the effects and control of shrinkage, the role of reinforcement, thermal properties, watertightness and the cause of fatigue. Also discussed will be the acoustical, electrical and elastic properties of concrete.

Lesson Notes: Expansion and contraction are important to the dimensional stability of the structure, and creep or plastic flow may cause an undesirable change in the stresses distributed through the structure. Once again, water is at the heart of most problems. As you are studying this chapter, see how factors such as shrinkage, bleeding and watertightness are directly or indirectly affected by the amount of water in the mix.

Key Points:

- At what point is concrete subject to shrinkage?
- Why does concrete shrink?
- Name the factors that affect shrinkage.
- Besides water loss, why else might concrete shrink?
- How can cement affect shrinkage?
- What is the most important factor in minimizing shrinkage?
- How would a water-reducing admixture affect shrinkage?
- Discuss the effects of aggregates on shrinkage.
- What is the maximum amount of clay that sand can contain before shrinkage is seriously affected?
- What percent of sand should pass a 100-mesh screen? A 50-mesh screen?
- What is the recommended slump for slabs?
- Identify five ways to control water loss.
- How is water lost from concrete?
- Define "plastic" shrinkage.
- What happens when there is a rapid loss of bleed water?
- Describe the effects of low humidity and wind.
- Can a minor change in weather have a great effect on evaporation? Explain using Figure 5-3.
- When is bleeding detrimental to concrete, and what are the negative effects?
- What is drying shrinkage?
- What has the greatest effect on drying shrinkage?
- How much shrinkage will occur with 300 pounds of water per cubic yard?
- What is the range of drying shrinkage?
- Name the factors that can help limit shrinkage.
- How is shrinkage estimated?
- How does the volume of concrete change when it gets warm or cool?
- What can happen when concrete is restrained from movement?
- Do volume changes caused by temperature affect concrete differently than those caused by moisture?
- How does reinforcement affect shrinkage?
- Describe the chemical methods of shrinkage control.
- Why should aluminum powder not be used to control shrinkage.
- When would the use of superplasticizers be appropriate.
- How could a volume change be measured?
- What is meant by the term "coefficient of expansion"?
- What variables can affect this?
- Why is thermal coefficient important?

- What occurs when there is no provision for movement?
- What is conductivity?
- Does concrete have a fairly high "k" value?
- Name the three things that influence concrete's conductivity.
- What is the Btu range for concrete?
- Identify the ways in which the "k" value of concrete is important.
- Define "specific heat" and "diffusivity."
- Define "modulus of elasticity."
- What is the stress-strain curve of hardened concrete?
- What might the elastic modulus tell us about concrete?
- How does fire affect the elastic modulus?
- What factors affect a reduction in elasticity?
- How is the modulus of elasticity related to compressive strength?
- Define "creep."
- What is the difference between creep and plastic flow?
- Over what period of time does creep continue?
- What is the rate of creep in relationship to time?
- Name the two components of creep.
- What process can reduce creep?
- In normal concrete mixes, how can creep be kept at a minimum?
- Define "permeability."
- On what does the permeability of concrete depend?
- How is porosity affected by the water-cement ratio?
- Is pressure required to create capillary flow through porous concrete?
- Identify three ways that water under a hydrostatic head can flow through concrete.
- Name the three factors that are most important to the watertightness of concrete.
- Explain why waterproofed cement and additives are not recommended.
- In most cases, of what is a lack of watertightness a result?
- List the six principles and precautions for obtaining watertightness.
- Describe two methods for minimizing moisture problems on enclosed slabs.
- Summarize the best way to obtain impermeable concrete.
- What type of concrete is best for sound control within and between areas?
- Is concrete an insulator or conductor?
- Define "fatigue limit."
- What factors can influence fatigue in concrete?
- Define "yield."
- Identify the factors that can contribute to loss of yield.

CHAPTER 5–QUIZZES

I Multiple Choice

1. Lack of watertightness in concrete can almost always be traced to _____.
   a. porous aggregates
   b. improper cement/aggregate proportions
   c. poor construction practices
   d. creep
   e. waterproofing admixtures
   Response _____ Reference _____

2. When used as an accelerator _____ causes an increase in shrinkage.
   a. pozzolan
   b. fly ash
   c. calcium chloride
   d. tricalcium aluminate
   e. sandstone
   Response _____ Reference _____

3. Which of the following does not affect shrinkage in concrete?
   a. water-cement ratio
   b. aggregate grading
   c. weather conditions
   d. cement volume
   e. quality of curing
   Response _____ Reference _____

4. Moisture problems associated with slabs on the ground can be minimized by
   _____. a. installing a vapor barrier
   b. laying a 1-inch sand base sub-base
   c. using an admixture that helps to retain water
   d. air-entrainment
   e. using less water in the mix design
   Response _____ Reference _____

5. The rate at which a material conducts heat through a 1-inch thickness per unit of area is known as _____.
   a. Btu
   b. diffusivity
   c. "k" value
   d. modulus
   e. coefficient of expansion
   Response _____ Reference _____

6. The property of concrete that indicates its ability to change in volume with changes in temperature is known as its _____.
   a. conductivity
   b. coefficient of expansion
   c. diffusivity
   d. modulus of elasticity
   e. dynamic creep
   Response _____ Reference _____

7. Which of the following factors will not help limit shrinkage in concrete?
   a. smallest size aggregate possible
   b. proper consolidation
   c. good workmanship
   d. proper curing
   e. intelligent use of admixtures
   Response _____ Reference _____

II True/False

8. Yield is defined as the volume of concrete per cubic yard.
   T ___✓___ F _____ Reference _____

9. The most critical factor for minimizing shrinkage in concrete is the total water per cubic yard.
   T ___✓___ F _____ Reference _____

10. Reinforcing steel is rarely used to help control shrinkage.
    T _____ F ___✓___ Reference _____

11. Concrete does not start losing water for about 15 to 20 minutes after placement unless Type III cement is used or concrete is in contact with earth.
    T _____ F ___✓___ Reference _____

12. Creep is a time-dependent deformation of concrete under varying loads.
    T _____ F ___✓___ Reference _____

13. Entrained air decreases drying shrinkage, but because air entrainment requires the use of more water, the effect on shrinkage is negligible.
    T _____ F ___✓___ Reference _____

14. A small amount of bleeding is not detrimental to concrete and, in fact, can result in a slightly stronger paste.
    T ___✓___ F _____ Reference _____

III Completion

15. _____LOW_____ humidity in the air and _____WIND_____ are the principle causes of high evaporation. However, _____AIR_____ temperature can also be significant.
Reference _____

16. Aluminum is not a good way to control shrinkage and should not be used in normal construction because of _____ and the possible _____ of strength.
Reference _____

17. Concrete is a _____ conductor of sound because it is a _____ material.
Reference _____

18. When water loss is fairly slow, the concrete can adjust to the reduction in _____, whereas a rapid loss of _____ water from the surface of a slab will introduce a _____ stress in the surface layer.
Reference _____

19. The modulus of elasticity is the _____ of a substance and is known by the letter _____.
Reference _____

20. Volume change is the _____ and _____ of concrete that results from temperature changes or _____ drying. These changes are _____.
Reference _____

CHAPTER 6
CRACKS AND BLEMISHES

Objectives: To become familiar with the causes and prevention of cracks and blemishes and to obtain an understanding of how repairs to concrete are made.

Lesson Notes: The properties of concrete are all interrelated. When one symptom appears, we can be sure that other properties will be affected. Cracks and blemishes seen on the surface usually indicate a problem below the surface that cannot be seen.

Key Points:

Cracking:
- Cracks and blemishes can result from a deficiency in which properties of concrete?
- Can cracking be prevented?
- Why does concrete crack?
- What are the main causes of cracking?
- What are plastic cracks?
- How quickly can plastic cracks develop?
- Describe the possible size, length and shape of plastic cracks.
- How do plastic cracks differ from cracks in hardened concrete?
- Where do plastic cracks usually occur?
- Describe how weather can influence cracking?
- How can plastic cracking be controlled?
- How does evaporation affect plastic cracking?
- Identify the ways that cracking can occur prior to hardening.
- Describe how settlement or movement in the concrete, forms, subgrade and soil can contribute to cracking.
- At what location may settlement occur in a monolithically placed beam and column or slab and wall combination?
- What is the cause of drying shrinkage cracks?
- What role does restraining of concrete play in cracking?
- Name the other important factors that contribute to drying shrinkage cracks.
- How do tensile force and tensile strength relate to cracking?
- What is a structural crack?
- How and where might structural cracks occur?
- What are the job conditions that can cause structural cracks?
- What acts of nature can cause or contribute to structural cracks?
- What is the result of reactive aggregates in hardened concrete?
- What factors contribute to reactive aggregates?
- Describe how rusting or reinforcing steel can cause cracks and how to prevent it.
- Define "thermal shock."
- How does it occur and what is the result?
- Where do weathering cracks occur most frequently?
- What will occur if these cracks are ignored?
- At what point do freezing and thawing cycles no longer affect concrete?
- Define "crazing."
- When is it most noticeable?
- Is crazing serious? Justify your answer.
- Identify the three general causes of crazing and why they occur.

Blemishes:
- Define "formed concrete."
- Describe the three general types of defects.
- What is meant by the term "dusting"?
- How can a dusting surface be made hard?
- When might a wood product cause dusting?
- Why is tannin harmful to concrete?
- In what way might heaters have a negative effect on plastic concrete?
- How can it be avoided?
- What is the most frequent cause of dusting?
- How does a lack of curing or dirty aggregate create dusting?
- What causes bugholes?
- Do bugholes create structurally unsound concrete?
- What are the ways to eliminate or reduce bugholes?
- Describe the process of sack rubbing.
- How is a concrete surface stoned, and how does it differ from sack rubbing?
- Name the causes of bubbles and blisters.
- What are rock pockets, and how do they form?
- What are the principle causes of rock pockets?
- How can you prevent concrete from sticking to forms?
- What conditions make concrete stick to metal forms?
- To what weight factors should forms be designed?
- What is a primary concern in designing forms?
- How might improper spacing of studs, walers and form ties affect forms?
- How might a blemish occur at a horizontal construction joint?
- What areas comprise the largest unformed surfaces?
- List the types of materials that may stain or discolor concrete.
- How should calcium chloride be incorporated into concrete.
- What is it about plywood that can stain concrete?
- When using white cement, what materials should be avoided?
- Why should dry cement NOT be used to absorb water?
- Name the possible causes for irregular dark areas in slabs.
- What procedures can be used to minimize dark spots in slabs?
- Define "efflorescence."
- What type of structure is most susceptible to efflorescence?
- How is efflorescence formed?
- How can efflorescence be reduced?
- Describe how efflorescence is removed.
- What precautions should be practiced when using acid to remove efflorescence?
- What is laitance?
- What are the properties of laitance?
- What are the causes?
- Define "scaling."
- What are the causes?
- Describe the work procedures that can cause scaling.
- Identity the best preventative measures for scaling where freezing and thawing cycles are present.
- Define "spalling."
- List the causes of spalling.
- How is spalling avoided?
- What is popout and what are the causes?

- What is usually present when popouts occur?
- How might cinder concrete suffer popouts?
- How are popouts prevented?
- Can popouts be repaired? If so, describe how.

Repair of Defects:
- What are the steps in repairing concrete?
- Name some causes of defects.
- Why make repairs to concrete?
- Describe the differences between structural and cosmetic repairs. Can a repair have both aspects?
- How much of the damaged or defective concrete must be removed?
- What are the methods used to repair concrete?
- Describe how the surface of the old concrete should look.
- What tools are used to repair concrete? Which are preferred?
- How is reinforcing exposed during concrete removal treated?
- After removal of defective concrete, what must occur?
- Do all patches require wetting of the old concrete?
- When is dry pack used?
- Of what materials and proportions does dry pack consist?
- Describe how dry pack is installed.
- When is it best to replace damaged concrete?
- Where might forms be necessary to repair concrete?
- Describe how concrete replacement for large areas is done.
- What is the procedure for repairing with an overlay?
- When might specialized procedures be used?
- When is crack repair useless?
- Is there significant gain from repairing cracking due to weathering?
- How should cracks be cleaned?
- What are the two general guides for repairing cracks?
- What types of materials can be used to fill cracks?
- How are large cracks filled?
- How can mortar bond be improved?
- What material works best for high strength and adhesion?
- Describe the epoxy process for filling cracks in both vertical and horizontal elements.
- How do you identify good proprietary materials for each specific project?
- How are bonding agents applied?
- How are posts, markers and metal ties bonded to concrete?
- Describe the process of joining concrete with adhesives.

CHAPTER 6–QUIZZES

I Multiple Choice

1. Which of the following is not a crack that occurs while concrete is still plastic?
   a. green
   b. plastic shrinkage
   c. pre-set
   d. drying shrinkage
   e. none of the above
   Response _____ Reference _____

2. Sudden changes in temperature that can stress concrete and cause cracks are called _____.
   a. reactive thermoset
   b. thermal shock
   c. frost action
   d. freezing and thawing cycles
   e. drying shrinkage
   Response _____ Reference _____

3. Joint dowels in slabs on the ground should be _____.
   a. coated with a lubricant
   b. perpendicular to the subgrade
   c. secured against slippage
   d. placed off center
   e. all of the above
   Response _____ Reference _____

4. A deposit of crystalline salts on hardened concrete brought by water and deposited on the concrete surface through evaporation is called _____.
   a. laitance
   b. spalling
   c. scaling
   d. efflorescence
   e. drying scale
   Response _____ Reference _____

5. The minimum thickness of a bonded overlay for slab repairs should not be less than _____.
   a. 1 inch
   b. 2 inches
   c. 3 inches
   d. $1^1/_2$ inches
   e. $2^1/_2$ inches
   Response _____ Reference _____

6. Unvented heaters used for heating an enclosure during cold weather will cause a reaction when _____ come(s) into contact with the surface of plastic concrete.
   a. hydrogen ions
   b. ferruginous concretions
   c. chloride salts
   d. silica
   e. carbon dioxide
   Response _____ Reference _____

7. A small cone-shaped piece of concrete with the base on the surface of the concrete is called a _____.
   a. scale
   b. spall
   c. popout
   d. pit
   e. void
   Response _____ Reference _____

8. _____ cracks are caused primarily because of rapid loss of water from new concrete after it has hardened.
   a. Plastic shrinkage
   b. Spalling
   c. Drying shrinkage
   d. Hydration
   e. Contraction
   Response _____ Reference _____

II True/False

9. Concrete in structures consisting of a large amount of concrete in huge blocks or masses is called mass concrete.
   T ✓ F _____ Reference _____

10. Contraction joints should be spaced not more than about 30 feet apart.
    T _____ F ✓ Reference _____

11. When concrete is first placed in forms it contains large amounts of entrapped air that cause voids called rock pockets, which can be removed if proper vibration is applied.
    T ✓ F _____ Reference _____

12. Discoloration of concrete can be caused by certain plywoods, hardboards, form oils and iron pyrites.
    T ✓ F _____ Reference _____

13. The dry pack method of repairing concrete has an advantage in that it requires special knowledge to use and can only be applied by certified installers.
    T _____ F ✓ Reference _____

14. Surface to be bonded by adhesives must be sound and thoroughly wetted prior to application.
T _____ F _✓_ Reference _____

15. Preparation for repair of concrete begins with removal of unsound and disintegrated concrete.
T _✓_ F _____ Reference _____

16. Concrete cracks are due to compressive forces that pull the concrete apart before tensile strength is adequate.
T _____ F _✓_ Reference _____

III Completion

17. Settlement of concrete may be obstructed by _____, _____ in the concrete or large_____ _____, causing _____ in the fresh concrete over these obstructions.
Reference _____

18. Diagonal cracks at corners of doors and window ___openings___ can be controlled by the use of sufficient _reinforcing_.
Reference _____

? 19. Isolation joints should be provided whenever concrete abuts_previously placed_ concrete in ___SLAB___, ___WALL___ or footings.
Reference _____

20. The first step in repairing concrete is to _DIAgnose_ the damage, including determination of the _cause_ and the _Extent_.
Reference _____

21. Cracking of precast concrete can be minimized if units are _Designed properly_, avoiding variable _Sections_ and providing adequate _reinforcing_.
Reference _____

22. _BuGHOLES_ are small pits, bubbles or voids about _1/2 in_ in diameter in formed concrete surfaces and are sometimes covered with a thin skin of _Dried paste_ that breaks under slight _pressure_.
Reference _____

23. One of the worst blemishes in a horizontal concrete structure is a sloughing away or _peeling_ of the surface in thin flakes called _Scaling_.
Reference _____

24. Large cracks can be filled with epoxy mortar consisting of epoxy _____
mixed with _____ in the proportion of _____ part _____
to _____ parts _____ by volume.
Reference _____

25. Often appearing as circular or oval depressions on surfaces, _____ is
a deeper surface defect than scaling that may be _____ or more in depth
and _____ or more in diameter.
Reference _____

CHAPTER 7
PORTLAND CEMENT

Objectives: To obtain a basic understanding of the way cement is manufactured; its composition, properties and characteristics; and the methods of its transportation and storage.

Lesson Notes: For a better understanding of how cement is made, study Figure 7-3 as you read Section 7.2.

Key Points:

- What is meant when it is said that cement is hydraulic in nature?
- Of what raw materials is cement made?
- What raw material makes up the largest proportion of cement?
- Besides limestone, name three other sources of carbonate rock that are used for making cement.
- Of how many materials might cement be composed?
- What is the process of making cement called?
- Describe the first phase of cement manufacture.
- After the blended material is stored, what are the two possible processes prior to its being sent to the kiln? Describe each.
- Describe the burning and finishing process.
- What are clinkers?
- What materials can be added during finish grinding?
- Name four primary and two minor compounds that compose most cement.
- What methods are used to determine cement particle size?
- Describe each of the five main types of cement, including the characteristics and uses of each.
- What are the three types of air-entrained cements available?
- What is blended cement?
- What is added to cement to make each of the following types? IS, IS-A, P, IP, S, I(SM) and I(PM)
- What is masonry cement?
- How does white cement differ from gray cement? Is there a structural difference between them?
- Name some uses of white cement.
- What is added to cement to make plastic cement, and what are its most common uses?
- What does waterproof cement contain?
- How does expansive cement differ from other cements?
- Where is expansive cement used most effectively?
- Calcium aluminate cement is used for what applications?
- Can aluminous cement be used for structural elements?
- How is magnesite made, and where is it used?
- Where is rapid-setting cement used most frequently?
- What special concerns are related to rapid-setting cements?
- Identify the ways in which the following properties and characteristics of cement affect concrete: color, finesses, soundness, setting time, compressive strength, heat of hydration, loss of ignition and specific gravity.
- Describe the methods for moving cement between site locations.
- What are the two basic types of hauling equipment used to transport cement?
- Cement must be protected from which natural elements?
- When being used, in what condition should cement be?
- What is warehouse set?
- Can brands and types of cement be stored together? Explain your answer.
- How is bulk cement usually stored?

- What problems are associated with cement stored in silos?
- What are two concerns when storing cement?
- Why is it important for all equipment used in handling cement to be weathertight?
- Describe how sacked cement should be stored at warehouses, stores and job sites.

CHAPTER 7–QUIZZES

I Multiple Choice

1. Generally speaking, the specific gravity of portland cement is about _____.
   a. 2.75
   b. 2.92
   c. 3.15
   d. 3.25
   e. 3.40
   Response _____ Reference _____

2. Type IV cement is a special cement that generates less heat during hydration and is used only in mass concrete such as _____.
   a. high-rise buildings
   b. large parking structures
   c. large dams
   d. tilt-up buildings
   e. water treatment plants
   Response _____ Reference _____

3. The process of making cement is called _____.
   a. hydration
   b. hydraulic kiln refining
   c. clinker
   d. pyroprocessing
   e. heat stearation
   Response _____ Reference _____

4. Type _____ cement is used where high strengths are desired at early periods.
   a. I
   b. II
   c. III
   d. IV
   e. V
   Response _____ Reference _____

5. Which of the following is not a property or characteristic of cement?
   a. fineness
   b. setting time
   c. color
   d. workability
   e. soundness
   Response _____ Reference _____

Concrete Manual Study Companion

II True/False

6. The divergence of dry and wet paths ends when the kiln feed is put into storage.
T ✓ F _____ Reference _____

7. Cement sacks can be stacked directly on a warehouse floor, provided there is no moisture coming through the floor.
T ✓ F _____ Reference _____

8. Shipments of cement to the customer are made either in bulk or in 90-pound bags, which equal about a $1/2$ cubic foot.
T _____ F ✓ Reference _____

9. Color is an indication of cement quality.
T _____ F ✓ Reference _____

10. One problem with storing cement in silos is a tendency to develop a hollow core in the center when the cement is withdrawn from the bottom.
T ✓ F _____ Reference _____

11. Slag cement (ASTM C 595) may be used as a cementing ingredient for structural concrete.
T _____ F ✓ Reference _____

III Completion

12. Type V cement is a special __sulfate resistant__ cement. It is used where concrete is exposed to __soil__ or __ground__ water that is high in __sulfate__ content.
Reference _____

13. Greater cement fineness increases the rate at which cement __Hydrates__ and __ACCELERANTe__ strength development.
Reference _____

14. Portland blast-furnace slag cement can be either Type __IS__ or __IS-A__, slag cement is Type __S__ and portland-pozzolan cement can be either Type __IP__ or __P__.
Reference _____

15. During finish grinding of cement, a small amount of __GYPSUM__ is interground with the cement to control __setting time__ of the cement when it is used.
Reference _____

16. White portland cement contains no _____ and meets the requirements for Type _____ portland cement. It is pure white in color and allows for a great amount of variety in _____ or _____ concrete.
Reference _____

40

CHAPTER 8
AGGREGATES

Objectives: To identify the different types and sources of rock used as aggregate, as well as the characteristics, processing, stockpiling and testing of aggregate materials. Also, the special kinds of aggregates will be studied in a brief overview.

Lesson Notes: Aggregates are normally inert materials and do not react with the concrete; however, there are some aggregates to which this generally does not apply. As you study this chapter, note the types of aggregates that may react with the concrete.

Key Points:

- What aggregates are used in concrete?
- How much of the volume of concrete is occupied by aggregates?
- How do most aggregates affect the physical or chemical reactions within concrete?
- What class of rock makes the most consistently good aggregate?
- How are the three rock classes formed?
- Describe the differences between the three rock classes.
- Where are most natural aggregates found?
- How might gravel deposits originate?
- Where do most glacial aggregates occur?
- Identify the main kinds of water-transported aggregate deposits.
- How do these deposits occur?
- What is the first step in obtaining aggregate approval?
- In efforts to obtain good aggregates, what problems can occur in rural or isolated areas? What are their resolutions?
- What purpose does testing and inspection play in obtaining good aggregate?
- How is aggregate quality determined?
- List the seven properties that affect aggregate quality.
- How are aggregate soundness and stability determined?
- Should aggregates be accepted or rejected based only on this test?
- How is cleanness determined?
- Name the materials that can negatively affect aggregate quality.
- Identify the potential effects of these materials on concrete.
- How is aggregate hardness determined?
- Should a hardness test be the sole determination of good-quality aggregate?
- By what name is a common grading test known?
- Describe how this test is performed and how its results are analyzed.
- How is the fineness modulus determined?
- What is the most desirable grading curve?
- When might aggregate grade be undesirable?
- Review Section 3.11 on the maximum size aggregate (MSA). What effect does the MSA have on concrete?
- In what shapes is aggregate found?
- What is the main influence on aggregate shape?
- What are the effects of aggregate shapes on concrete hardness, flexural strength, compressive strength and workability?
- Describe the differences between aggregate shape and texture.
- Which texture is most desirable?
- Why is a petrographic analysis of aggregate important?

- Define "specific gravity."
- How can the specific gravity of aggregate affect concrete?
- What might a low specific gravity indicate?
- Does low specific gravity determine aggregate acceptability? Explain your answer.
- Describe how absorption affects aggregate quality.
- Why must the absorption of an aggregate be known?
- Identify the four possible moisture content conditions.
- Why is knowing the moisture content necessary?
- Define "unit weight."
- What is void content, and why is it important?
- Study Table 8.7. Why is it rare to find aggregate that is dug out of the ground ready to be used in concrete?
- How is poor grading remedied?
- What is the first step in processing aggregate?
- Evacuation of materials is done by what means?
- In what conditions should sand and gravel be processed?
- How much water might be used for washing aggregate?
- What are some concerns associated with wash water?
- What should be removed from coarse aggregate before primary crushing?
- What equipment is used for initial, intermediate and final crushing?
- Describe the purposes of a revolving scrubber, a log washer and a screw washer.
- Define "fine aggregate."
- How is sand grading accomplished?
- Review the effect of sand grading on concrete.
- How can the defects of pit-run sand be corrected?
- What are two alternatives when correcting sand grading using blended sand?
- Why is hydraulic separation best for sand sizes?
- Compare the different hydraulic classifiers.
- What is aggregate beneficiation?
- Compare and contract the four most common processes for improving aggregate.
- What are the two types of quarries?
- Briefly describe each type.
- What are the steps required to maintain aggregate stockpiles?
- Name the three types of equipment used to remove material from a stockpile.
- State how good mixing is achieved using each equipment type.
- How can segregation be minimized when stockpiling coarse aggregate?
- How does sand differ from coarse aggregate in regard to segregation?
- What standard is usually specified for aggregate sampling and testing?
- How is moisture in sand usually measured?
- What standard should be followed when sampling aggregate?
- Why is sampling from a stockpile difficult?
- Give an example of a sampling plan.
- How should a sample be obtained from a conveyor belt?
- If possible, when should sampling be avoided?
- How should samples be sent to the laboratory?
- When is the quartering method used for sampling?
- Describe this method.
- When should testing be done?
- What is slag?
- How is slag processed?
- How does slag compare to natural aggregate?

CHAPTER 8–QUIZZES

I Multiple Choice

1. The greater size range covered by the gravel in a pile, the greater the danger of harmful _____.
   a. beneficiation
   b. hydration
   c. segregation
   d. scrubbing
   e. rounding
   Response _____ Reference _____

2. Which of the following is not a characteristic of an aggregate?
   a. cleanness
   b. durability
   c. texture
   d. reactivity
   e. absorption
   Response _____ Reference _____

3. When taking a sample of sand for testing, the sample size should be _____ pounds.
   a. 10
   b. 20
   c. 30
   d. 40
   e. 50
   Response _____ Reference _____

4. The limit of deleterious substances in aggregate should not be more than _____ percent by weight, depending on the substance.
   a. one to two
   b. two to three
   c. three to four
   d. four to five
   e. five to eight
   Response _____ Reference _____

5. The particle shape of an aggregate that will tend to make a harsh concrete is _____.
   a. angular
   b. rounded
   c. subrounded
   d. crushed
   e. circular
   Response _____ Reference _____

6. Unsatisfactory grading of aggregates can be corrected by _____.
   a. breakage
   b. segregating
   c. crushing and screening
   d. scalping
   e. spalling
   Response _____ Reference _____

7. Coarse aggregate sample should be reduced by using _____.
   a. a sample splitter
   b. the quartering method
   c. dry selection
   d. wet selection
   e. beneficiation
   Response _____ Reference _____

8. Sand or fine aggregate for concrete consists of material that will pass a No. _____ screen.
   a. 4
   b. 5
   c. 6
   d. 7
   e. 8
   Response _____ Reference _____

II True/False

9. A number that is useful when studying aggregate gradation is the fineness modulus.
   T ✓ F _____ Reference _____

10. Aggregates in concrete are frequently called filler material because they occupy between 60 and 80 percent of the volume.
    T ✓ F _____ Reference _____

11. The quality of rock in a quarry is fairly consistent, especially the limestone and granite.
    T _____ F ✓ Reference _____

12. Aggregates for structural concrete can be either natural or artificial and may weigh as little as 75 pounds per cubic foot.
    T _____ F ✓ Reference _____

13. Natural aggregates used in concrete come either from solid bedrock or deposits of sand and gravel.
    T ✓ F _I_ Reference _____

14. Sand and gravel are most frequently dug out of the ground and used directly in concrete.
T _____ F ✓ Reference _____

15. Segregation of materials in a gravel stockpile can be minimized by having a greater size range.
T _____ F ✓ Reference _____

16. The average specific gravity of sand or gravel is 2.65, which means it is 2.65 times as heavy as water.
T ✓ F _____ Reference _____

III Completion

17. Aggregate samples are taken from a conveyor belt by stopping the belt, taking at least _____ portions and combining them to form a sample. _____ templates shaped to fit across the belt must be inserted and all material between the templates, including _____ and _____, removed.
Reference _____

18. Of the several varieties, slag that comes from a _____ is the most suitable for use in concrete.
Reference _____

19. There are _____ basic classes of rock. They are _____, _____ and _____.
Reference _____

20. Hard, dense stone such as granite may have an absorption rate of only _____ percent, whereas the absorption rate of a shale or porous chert is as high as _____ percent. The absorption rate for sand should not exceed _____ percent.
Reference _____

21. The limits for deleterious substances in fine aggregate for concrete is about _____ percent for clay lumps and between _____ and _____ percent for coal and lignite.
Reference _____

22. There are four commonly used methods of beneficiation. They are _____ separation, _____, _____ and _____.
Reference _____

23. Scrubbing is required when adherent coatings of _____ and _____ cannot be removed from aggregate by washing and screening. The three methods of scrubbing are use of a _____ scrubber, a _____ or a _____.
Reference _____

24. To avoid segregation of materials when stockpiling aggregate, the following precautions should be observed. Handle as _____ times as possible, avoid _____, _____ shaped piles, stockpile in _____, handle in _____ graded sizes and remove from the stockpile in _____ slices.
Reference _____

25. A _____ texture is desirable in aggregates as it provides better bond with the _____, making concrete of better strength compared with _____ surfaced aggregates.
Reference _____

CHAPTER 9
WATER AND ADMIXTURES

Objectives: To understand the effects of water, various admixtures, pozzolans and fly ash on plastic, fresh and hardened concrete.

Lesson Notes: Water is absolutely necessary. It lubricates and makes concrete plastic and workable, as well as provides the catalyst for the reaction with the cement. However, when the amount of water exceeds the specified limits, the benefits of water become liabilities. As the water-cement ratio rises, strength, durability, workability and other properties of concrete diminish.

When used, admixtures must conform to American Society for Testing and Materials (ASTM) standards and the manufacturer's specifications, and care must be taken to strictly adhere to them. Before introduction into a mix, the effects of an admixture on the other concrete materials and the site conditions must be known, or defects in the concrete may result.

Key Points:

- Name two things that water does to cement.
- What is the water-cement ratio for good workable concrete?
- Why is increasing the water-cement ratio not good for concrete?
- Describe the ideal water for concrete.
- Up to how much dirt or silt is acceptable for water used for concrete?
- Define "ppm" and "TDS."
- In what ways might sugar in the water affect concrete?
- Where is water contamination least likely?
- Without testing, how can contaminated water be identified?
- Describe the possible effects of sea water if used in concrete.
- What effect does sea water have on reinforcement?
- What are the three general classes of admixtures and the seven types of chemical admixtures?
- Name some concerns when choosing an admixture.
- What effects need to be known when using admixtures?
- Does every admixture affect all of concrete's properties? Explain your answer.
- How might an admixture have variable effects on concrete?
- Why use admixtures?
- What possible effects are sought when using admixtures?
- How should an admixture be tested?
- What three concerns should be kept in mind when selecting an admixture?
- How are liquid admixtures measured?
- When using an admixture, what capabilities should the dispensing system have?
- Why should an admixture in a dry or powdered state never be introduced into concrete?
- Are there any exceptions?
- Name some methods for dispensing admixtures.
- Should admixtures be intermixed prior to mixing? Why or why not?
- Is the time frame for adding admixtures ever critical? Explain.
- What does an accelerator do to concrete?
- What are the benefits of an accelerator?
- How and in what manner should calcium chloride be added to concrete?
- Identify the effects of calcium chloride on plastic, fresh and hardened concrete.
- How does a water reducer affect concrete?
- Name the most common chemical used for water reduction.

- What are the advantages of water reducers?
- How does a retarder affect concrete?
- Which chemicals make the best retarders?
- Define what is meant by "vibration limit"?
- How is a retarder evaluated?
- How might temperature affect a retarder?
- Describe the differences between a lignosulfonate and a hydroxylated retarder.
- Where might air-entraining agents be most frequently used?
- Name the most common-air entraining agents.
- Identify the benefits of air-entrainment.
- Describe how the disadvantages of air-entrainment can be offset.
- When might adding alcohol be necessary?
- When is the best time to add air-entraining agents to the concrete mix?
- Study Table 9.3. What factors can change the amount of entrained air?
- Name the most frequent causes of water leakage through concrete.
- When is a powdered waterproofing admixture of value?
- How do stearates, soaps and fatty acids affect concrete's waterproof qualities?
- Identify the types of bonding agents most frequently used for concrete.
- What type(s) of compounds can be used as antifreeze agents?
- What requirements should coloring admixtures meet?
- Is the use of trial panels recommended? Why or why not?
- What are the best workability agents?
- What materials make up expansive cement?
- How can shrinkage be chemically treated?
- Name the four kinds of finely divided mineral admixtures.
- When can a superplasticizer be used in concrete?
- How does a superplasticizer react with concrete?
- What benefits can be obtained from using a superplasticizer?
- Define "pozzolan."
- Name the three general classes of pozzolans and of what they are composed.
- Fully describe the reaction and general effects of pozzolans on concrete.
- Where are pozzolans most frequently used?
- What are the two classes of fly ash and how are they produced?
- What are the benefits of including fly ash in a concrete mix?
- Describe the density, size, color and form of silica fume. Of what is it a by-product?

CHAPTER 9–QUIZZES

I Multiple Choice

1. A retarder is an admixture that _____ the chemical process of hydration.
   a. increases
   b. slows
   c. accelerates
   d. stops
   e. none of the above
   Response _____ Reference _____

2. The two basic classes of fly ash are _____.
   a. A and B
   b. B and D
   c. C and F
   d. D and G
   e. A and D
   Response _____ Reference _____

3. The reason for using an admixture is to _____ of concrete so it will be more suitable for a certain usage.
   a. change the slump
   b. reduce segregation
   c. enhance the chemical properties
   d. reduce the cracking
   e. modify the properties
   Response _____ Reference _____

4. A superplasticizer can be used in concrete to _____.
   a. reduce the amount of water
   b. reduce cement content without reducing strength
   c. produce a flowing, self-leveling concrete
   d. all of the above
   Response _____ Reference _____

5. Pozzolans usually improve _____ and reduce _____.
   a. plasticity, air-entrainment
   b. durability, water volume
   c. workability, bleeding
   d. hydration, heat loss
   e. drying shrinkage, cohesiveness
   Response _____ Reference _____

6. The total air content for frost-resistant concrete in a moderate exposure when the MSA is $^3/_4$ inch should be _____ percent.
   a. six
   b. five
   c. four
   d. four and one half
   e. three
   Response _____ Reference _____

7. Which of the following is an acceptable source for concrete mixing water?
   a. a private well
   b. the sea water
   c. a stagnant pool
   d. a brackish body of water
   e. a swamp
   Response _____ Reference _____

II True/False

8. An accelerator speeds up setting time and increases the rate of early strength development.
   T _✓_ F _____ Reference _____

9. Entrained air in concrete increases bleeding and reduces segregation tendencies.
   T _____ F _✓_ Reference _____

10. There is no material that can be put into a batch of concrete to lower the freezing point of the fresh concrete without damaging the concrete.
    T _✓_ F _____ Reference _____

11. Silica fume is a material that is used as a pozzolanic admixture.
    T _____ F _____ Reference _____

12. One source of pozzolans is calcined or burnt shales and slates heated in a stationary kiln and crushed and ground after cooling.
    T _____ F _____ Reference _____

13. In general, sugar in mixing water is not objectionable.
    T _____ F _✓_ Reference _____

14. Expansion-producing admixtures compensate for drying shrinkage of the concrete and are usually incorporated in expansive cement.
    T _✓_ F _____ Reference _____

III Completion

15. When using more than one admixture, they should not be _____ prior to introduction into the mixer unless the _____ state that it is permissible.
   Reference _____

16. Some of the most common natural pozzolans are _____, _____, _____ and _____.
   Reference _____

17. Bonding agents can be applied to the _____ to be bonded or used as admixtures, and can be made from _____ or _____ rubber or _____.
   Reference _____

18. Stearates, used as a permeability-reducing admixture, reduce _____ and retard _____ but are of little or no value if the water is under pressure.
   Reference _____

19. Dependent on other material costs, fly ash is more economical, as well as being more resistant to _____ and _____ reaction, and has a _____ of hydration.
   Reference _____

20. Coloring admixtures should be _____ in sunlight, _____ in the presence of alkalies, and have no adverse effects on _____ or _____ development.
   Reference _____

CHAPTER 10
ACCESSORY MATERIALS

Objectives: To understand the use, purpose and installation of sealants, resins, bonding agents and other coatings.

Lesson Notes: New materials are continually being introduced. It is important that these materials be tested for their intended use prior to installation. An untested product can quickly become a detriment to what otherwise would be good quality concrete.

Key Points:

- Why are construction joints needed?
- Name the various kinds of field-molded sealants.
- Describe each of the following sealants as to composition and where used: mastics, hot-applied thermoplastics, chemically curing thermosetting sealants, solvent-release thermosetting sealants and rigid materials.
- What are preformed sealants?
- Of what types of materials do they consist?
- Name the three types, grades and classes of epoxy resin systems.
- Where are epoxy resin systems used?
- What are the temperature ranges, conditions, surfaces and applications of epoxy resin systems?
- What two components are usually part of an epoxy resin system?
- What is the pot life of these systems?
- Name the advantages of bonding agents.
- What are bonding agents usually classified as?
- Identify the various means of coloring concrete.
- Describe the types of paints that can be used to improve durability, make watertight and decorate concrete.
- What materials can be used for waterproofing and dampproofing concrete?
- Why should plaster of paris not be used as a patching compound?
- Where would a surface retarder be used?

CHAPTER 10–QUIZZES

I Multiple Choice

1. _____ are thick liquids used where small joint movement is expected.
   a. Mastics
   b. Solvent-release thermosetting sealants
   c. Epoxies
   d. Thermoplastics
   e. Patching compounds
   Response _____ Reference _____

2. Epoxy resins will not normally adhere to _____ surfaces.
   a. wet
   b. metal
   c. wood
   d. concrete
   e. greased
   Response _____ Reference _____

3. Some rapid-setting cements contain _____, which causes a set within a few minutes.
   a. dehydrated gypsum
   b. hydrated lime
   c. epoxy resin
   d. mastic
   e. calcium chloride
   Response _____ Reference _____

4. Which of the following materials is not a chemically-curing thermosetting sealant?
   a. polysulfide
   b. epoxy
   c. urethane
   d. silicone
   e. neoprene
   Response _____ Reference _____

5. Which of the following is not an application in which epoxy resins are normally used with concrete?
   a. producing a skid-resistant surface
   b. bonding hardened concrete to other materials
   c. waterproofing and waterstops
   d. bonding plastic concrete to hardened concrete
   e. filling cracks
   Response _____ Reference _____

II True/False

6. Polyvinyl acetate, which improves the bond of concrete to old concrete, is a type of epoxy resin.
   T _____ F _____ Reference _____

7. A job-mixed paint to make concrete watertight is composed mainly of white portland cement and calcium stearate.
   T _____ F _____ Reference _____

8. One method of exposing aggregate on the surface of concrete is to use a surface retarder.
   T ___✓___ F _____ Reference _____

III Completion

9. Two methods of installing preformed sealants are to _____ the sealant in the concrete or install by _____ the sealant into the joint slot.
   Reference _____

10. Sealants that are cured by a release of a solvent include certain _____,
   _____, _____ and _____.
   Reference _____

11. Epoxy resins are usually composed of two components, the basic _____ and a _____.
   Reference _____

12. Rigid waterstops are usually made of _____; flexible waterstops are usually made of natural and synthetic _____ and _____.
   Reference _____

CHAPTER 11
FORMWORK

Objective: To gain an understanding of the various materials used for forms and the requirements for formwork including bracing, shoring, form oils, cleanliness and removal.

Lesson Notes: It is all too frequent that failure due to inadequate formwork causes major loss of life or property. Not included in the latter are the unsightly conditions that occur when only part of a formwork is deficient. There is no substitute for well-designed forms.

Key Points:

- Why are forms necessary?
- What are the various types of forms?
- Forms are required to be designed and built to resist what effects and factors?
- Name the 18 most common deficiencies that lead to the failure of forms.
- What are the types of materials of which forms can be made, and what is the most common?
- What are water and tie rods?
- The first consideration in designing formwork should be what?
- How could unsatisfactory alignment and vibration affect forms?
- Describe how form movement can be prevented.
- How should forms be constructed?
- When are chamfer strips used?
- Name the things of which wood forms should be free.
- When is plywood backing usually not necessary?
- In what dimension is plywood strongest?
- What should be the condition of metal forms?
- How should tie rods and metal ties be placed?
- Are openings entirely through the concrete usually permitted?
- Study Figure 11-9. Describe how horizontal construction joints should be formed.
- Why camber forms?
- Who normally chooses form materials?
- Name the most common wood species used for form lumber.
- What is the most stable type of lumber for forms?
- Why not use green or kiln-dried lumber?
- What are the most common uses for plywood in forms?
- Describe the construction of plywood sheets, the number of plys and the standard sizes.
- What is coated plywood? What are the available grades?
- In formwork, what is the most common use of glass fiber-reinforced plastic?
- Briefly describe how a fiberglass mold is made.
- What are the advantages of using plastic and rubber liners?
- What are the most common uses for steel forms?
- Of what are sonotube fiber forms made?
- Describe waste molds, their uses and the precautions necessary for good concrete.
- What is the most common form fastener?
- Describe each of the following and how they are used: form clamp, snap tie, coil tie, she-bolt and inserts.
- Why treat forms with oil?
- Name the different types of materials used as form coatings.
- How should coatings be chosen?
- Name the two general classes of form coatings and their subdivisions.

- How are the chemically active coatings applied to forms?
- Define falsework, permanent shores and reshores.
- What criteria should govern the installation of reshores?
- Define "slipform."
- For what are precast concrete forms used?
- How is bonding accomplished in precast forms?
- Identify the two types of prefabricated forms and the materials of which they are made.
- Prior to placing concrete, what should be done to forms?
- Where should runways be placed?
- What are the concerns related to metal chairs?
- What are the benefits of careful form removal after placing concrete?
- Why clean forms after use?
- When can forms be removed?

CHAPTER 11–QUIZZES

I Multiple Choice

1. The most common material used for forms is __b__.
    a. steel
    b. wood
    c. masonry
    d. hardboard
    e. sonotube
    Response _____ Reference _____

2. A __d__ is made from multiple layers of heavy paper bonded together and impregnated with resin or wax to become water-repellent.
    a. slipform
    b. hardboard form
    c. flexible liner
    d. sonotube
    e. waste mold
    Response _____ Reference _____

3. Prefabricated forms that can be used for many applications are known as __a__ forms.
    a. modular
    b. job specific
    c. slip
    d. spreader
    e. chamfer
    Response _____ Reference _____

4. Which of the following is not used as a form oil or compound? __d__
    a. wax
    b. lacquer
    c. plastic coatings
    d. motor oil
    e. shellac
    Response _____ Reference _____

5. A __d__ is a movable form that is raised vertically as the concrete is placed.
    a. roller form
    b. reshore form
    c. self-adjusting form
    d. slipform
    e. none of the above
    Response _____ Reference _____

6. An assembly for a wall form that is comprised of two nut washers, two waler rods and a central tie is a __C__.
   a. form clamp
   b. snap tie
   c. she-bolt
   d. coil tie
   e. reshore tie
   Response _____ Reference _____

7. Overlay plywood can be used without __b__.
   a. walers
   b. form oil
   c. chamfers
   d. bulkheads
   e. resin
   Response _____ Reference _____

8. __b__ should be placed in the corners of forms to produce beveled edges on permanently exposed concrete surfaces.
   a. edge protectors
   b. chamfer strips
   c. steel liners
   d. form clamps
   e. walers
   Response _____ Reference _____

9. Vertical shoring under a beam or slab can be accomplished either with permanent shores or __C__.
   a. slipshores
   b. precast shores
   c. reshores
   d. waste shores
   e. panel shores
   Response _____ Reference _____

II True/False

10. A snap tie is made of a single piece of wire cut to length and headed at each end.
    T __✓__ F _____ Reference _____

11. A waste mold is usually made of casting plaster reinforced with fiber and supported on wood framework.
    T __✓__ F _____ Reference _____

12. Clamps and pins should hold forms rigidly together in place and allow removal without damage to the concrete.
    T __✓__ F _____ Reference _____

13. Forms should be constructed to withstand the hydraulic head of at least 250 pounds per lineal foot.
    T _____ F __✓__ Reference _____

14. The quality of lumber that is usually specified to be used for formwork is utility grade.
    T _____ F __✓__ Reference _____

15. Except for prefabricated forms, forms usually are not designed for reuse.
    T _____ F __✓__ Reference _____

III Completion

16. Forms for suspended slabs and beams are frequently cambered to allow for __sagging__ or __settlement__, a common allowance being __1/4"__ per 16 feet of __span__.
    Reference _____

17. When placing a successive lift of concrete on previously placed and hardened concrete, the horizontal __joint__ between the two lifts is often a source of disfigurement that can be avoided by providing form __anchorages__ about __4 inches__ below the top of the __lift__.
    Reference _____

18. After stripping a form, it should have all __dirt__, __mortor__, __hardware__, and __other materials__ removed before reuse.
    Reference _____

19. When made, waste molds should be sized with __shellac__ or __lacquer__ and coated with parting compound or __form oil__ just prior to placing concrete.
    Reference _____

20. Prefabricated forms are held together with __locking devices__ and can be __joined together__ and __stacked__ to form large areas.
    Reference _____

CHAPTER 12
PROPORTIONING THE CONCRETE MIXTURE

Objectives: To understand how to proportion materials in a concrete mixture and how to adjust the mix to maintain the required quality, to review the properties of materials and understand the selection of mix characteristics when tests or history are not available.

Lesson Notes: Give additional attention to the steps used to estimate mix proportions in Section 12.3.

Key Points:

- Study Table 12.1, and note how changing the MSA affects a concrete mix.
- What are the basic requirements for concrete established by the type of structure and exposure conditions?
- How are the ingredients of concrete selected?
- In good quality concrete, what areas does the past occupy?
- When the quality of the cement paste is constant, what can be varied in a concrete mix?
- What is the result of wise material selection on a concrete mix? What qualities does the mix have?
- How should mixes be proportioned?
- Why should a mix be adjustable?
- What specifications may be included in a mix?
- According to the ACI 318 Standard, what must concrete possess?
- Regardless of the mix selected, what are the special exposure requirements that may have to be met?
- Identify the alternate methods that the code provides for selecting mix proportions.
- Define the following terms: "specific gravity," "bulk specific gravity," "density," "voids," "unit weight" and "absolute volume."
- What are the average weights of one cubic foot of cement, aggregate and water?
- List the steps in establishing a trial mix.
- What are the limits of the MSA?
- What is the most common aggregate size for structural concrete?
- Name the controlling conditions when water-cement ratio is not specified.
- What does slump measure?
- On what does the total amount of mixing water for the required slump depend?
- Using the percentage method, how are the coarse and fine aggregates determined?
- Review the mix design example given on page 138 of the *Concrete Manual*.
- When using trial mixes, how many mixes are to be made in order to establish strength versus water-cement ratio?
- Describe the final adjustments to be made to a proposed mix.
- Identify the differences when using the ACI method.
- Review the example on page 141 of the *Concrete Manual*. What is the variable when using air entrainment?
- What variables are involved when using superplasticizers and fly ash. Review Section 9.5.
- What are relative and actual yield?
- Identify the two small job alternatives.
- How is bank run sand and gravel to be used? Describe how the mix proportions are selected and measured.
- What information should be supplied when ordering ready-mixed concrete?
- What are gap-graded mixes?
- Where would gap-graded mixes be used?
- What might be the advantages of using gap-graded mixes?

CHAPTER 12–QUIZZES

I Multiple Choice

1. When the MSA in nonair-entrained concrete is one inch, the air content is 1.5 percent and the volume of sand is 41 percent, the amount of water per cubic yard will be approximately _____ pounds.
   a. 280
   b. 300
   c. 325
   d. 340
   e. 355
   Response _____ Reference _____

2. Relative yield is the _____ divided by the designed sign of the batch.
   a. unit weight
   b. actual yield
   c. weight of all materials except admixtures
   d. water-cement ratio
   e. aggregate weight
   Response _____ Reference _____

3. All aggregate particles in concrete should be _____.
   a. surrounded by paste
   b. moist prior to mixing
   c. added to the mix last
   d. clean, dry and segregated
   e. none of the above
   Response _____ Reference _____

4. Most aggregate is graded from the finest material to the MSA; in _____ grading, some sizes of aggregate are not used.
   a. gap
   b. batch
   c. selective
   d. stepped
   e. none of the above
   Response _____ Reference _____

5. If enough test results are not available for a statistical analysis of field tests, then it is necessary to make _____ to determine the concrete proportions.
   a. structural models
   b. strength value models
   c. a field analysis
   d. educated guesses
   e. laboratory trial batches
   Response _____ Reference _____

6. Which of the following material properties is not important when determining mix proportions?
   a. density
   b. unit weight
   c. voids
   (d) durability
   e. specific gravity
   Response _____ Reference _____

7. When the water-cement ratio is constant and sources of ingredients differ, concrete strengths _____.
   a. are constant
   b. are higher
   c. do not vary
   d. are usually lower
   (e) may vary
   Response _____ Reference _____

8. The project specifications for a small concrete job indicate a compressive strength of 2500 psi; without trial mix data, the required water-cement ratio for this concrete should be based on _____ psi.
   a. 2500
   b. 3000
   (c) 3500
   d. 3700
   e. 4200
   Response _____ Reference _____

9. Non-air-entrained concrete with a maximum size aggregate of $^3/_4$ inch will have about _____ percent air content.
   a. 1.0
   b. 1.5
   (c) 2.0
   d. 2.5
   e. 3.0
   Response _____ Reference _____

II True/False

10. The introduction of a high-range water reducer has not presented another variable in proportioning mixes.
    T _____ F __✓__ Reference _____

11. Each sack of ready-sacked concrete mix contains cement, fine aggregate and coarse aggregate weighing either 60 or 100 pounds and is ready to use by adding water.
    T __✓__ F _____ Reference _____

12. The normal procedures for mix proportioning can, in general, be applied when using no-slump concrete.
    T ✓   F _____   Reference _____

13. Superplasticizers cannot be used successfully with fly ash.
    T _____   F ✓   Reference _____

14. The ratio of the weight of a piece of aggregate of 1 cubic foot volume to the weight of 1 cubic foot of water is the bulk specific gravity.
    T _____   F ✓   Reference _____

15. Two methods of arriving at the mix proportions for a job are the statistical method and laboratory trial batches.
    T ✓   F _____   Reference _____

16. The total amount of mixing water per cubic yard of concrete is significantly affected by the cement content but is not affected by temperature.
    T _____   F ✓   Reference _____

III Completion

17. When test and history information is not available, estimates of mix proportions can be determined by following a _series_ of steps, the first of which is to select the _MSA_ from the specifications or based on the _JOB_ conditions.
    Reference _____

18. There are two kinds of voids, those _inside_ the aggregate and those that are _between_ aggregate particles.
    Reference _____

19. The MSA of any mix should not exceed one third the _depth_ of a slab, _3/4_ of the minimum clear reinforcement spacing or between reinforcing and the _forms_, nor _1/5_ the narrowest dimension between form sides.
    Reference _____

20. When selecting a mix using the aggregate content percentage method, _all_ batches should have compressive strength tested with _2_ cylinders at _7_, _14_ and _28_ days.
    Reference _____

21. When selecting a mix using the ACI method, the computed weights are based on _dry_ ingredients.
    Reference _____

22. When proportioning mixes, the effects of admixtures on _water content_ and _slump_ of concrete must be taken into consideration.
    Reference _____

CHAPTER 13
TESTING AND CONTROLLING THE CONCRETE

Objectives: To understand why testing is needed, the types of tests conducted on fresh concrete and how they are taken, the curing and testing of strength cylinders, methods of rapid strength gain, and the sampling and testing methods used on hardened concrete.

Lesson Notes: Although extensive knowledge of testing procedures for concrete is not necessary for everyone, the basics should be known so that we understand the importance and objectives of testing.

Key Points:

- What adjustments to a mix might be necessary under field conditions?
- What factors may cause the need to adjust the amount of water in a mix to maintain a consistent slump?
- What is the cause and correction when a change in workability due to increasing harshness of a mix occurs during placement of concrete?
- How is control of concrete accomplished?
- Define "sample."
- What is the basic requirement when sampling concrete?
- What occurs when a sample is not representative of the concrete?
- Describe the criteria for proper test performance.
- What is the purpose of laboratory and field testing?
- Why are tests necessary?
- Why take more than one test?
- One page 148 of the *Concrete Manual* there are two statements, "Testing is a precise operation" and "An improperly made test is worse than no test at all." In your own words, state why these statements are true.
- When might a nonstandard test be appropriate?
- There are two groups of tests for concrete. Identify the tests that belong to each group.
- List the ASTM number and title of each of the tests that may be used in concrete testing.
- Which tests are normally performed in a laboratory?
- Describe the method for obtaining samples of fresh concrete.
- How often should these samples be taken?
- What does a slump test determine?
- Describe the form and dimensions of a slump cone.
- How is slump measured?
- List the steps in taking a slump test.
- What is a "Kelly Ball?" Describe its form.
- Describe the basic procedure for this test.
- How does this compare to a slump test?
- What is the criteria for the test location?
- Why take the temperature of fresh concrete?
- Where can the temperature test be made?
- Where are the limits and tolerances for entrained air found?
- When should air content tests be taken?
- Name the two types of air meters.
- Briefly describe how each of the meters function and the factors that may affect them.
- Which types of meters require a correction factor?
- What is the main source of errors in these meters, and how can errors be avoided?
- Should the same concrete used in one test be used in other tests?

- What is the purpose of a unit weight test?
- Describe the unit weight test procedure for fresh concrete.
- At what other time can unit weight be determined?
- What tests may comprise a mixer performance test?
- On what is evaluation based?
- What have tests on stationary mixers shown?
- How are samples for a mixer performance test obtained?
- Describe the procedure for obtaining a sample of uniformity of concrete produced in a truck mixer.
- What information does a concrete strength specimen provide?
- What factors might affect a strength specimen negatively?
- What is the most common size of a specimen cylinder?
- How often are specimens made? Read ACI 318, Section 5.6.2.
- Give a detailed description of the procedure for making a strength cylinder specimen.
- After making strength specimens, how are the cylinders handled?
- Under what conditions can specimens be stored at a job site?
- What might the dangers be if stored on the job site?
- Why might job site curing be done?
- How are specimens stored in the laboratory?
- How are most cylinders prepared for testing?
- Describe the precautions to be observed when the cylinder is capped.
- How is the strength test conducted?
- When a specimen has a low strength, what visual observations might give an indication of the cause?
- Identify some of the methods of measuring strength gain?
- What do all these tests have in common in relationship to the 28-day strength?
- Briefly describe the three ASTM methods of accelerated strength testing.
- What factors may effect early strength testing methods?
- State the various ways by which control of concrete can be estimated at the time of placement.
- Where might tests for flexural strength be desirable?
- How is flexural strength determined, and what is the standard size of specimens?
- Describe the process of obtaining a sample for flexural strength tests.
- What are the differences between tests for compressive and splitting tensile strength?
- When are tests made on hardened concrete?
- Name each of these types of tests.
- What is the most common method of sampling hardened concrete?
- Describe the aspects and concerns when taking core samples.
- For what other purposes might core samples be taken?
- After samples of hardened concrete are obtained, how should they be treated and handled?
- What does ASTM C-42 recommend prior to testing samples for strength?
- When might a correction factor need to be applied?
- How are cores dressed?
- Name the two methods of testing concrete "in-place."
- How is a Swiss hammer calibrated?
- In what ways is the accuracy of a Swiss hammer affected?
- When using a Swiss hammer, how many readings are taken?
- Briefly describe how a Windsor probe works.

CHAPTER 13–QUIZZES

I Multiple Choice

1. The ASTM specification for making and curing concrete test specimens in the field is number _____.
    a. C-39
    b. C-138
    c. C-143
    d. C-172
    e. C-173
    Response _____ Reference _____

2. In general, when total water in a mix is increased by _____ percent the slump will increase about 1 inch.
    a. one
    b. five
    c. two
    d. ten
    e. three
    Reference _____

3. Which of the following is not included as a mixer performance test for uniformity of concrete?
    a. slump
    b. unit weight
    c. strength
    d. air content
    e. none of the above
    Response _____ Reference _____

4. In order to properly perform a test, it is necessary to _____.
    a. follow standard methods
    b. have it be performed by qualified persons
    c. properly interpret the results
    d. use the proper equipment
    e. all of the above
    Response _____ Reference _____

5. The most common size of a compressive strength cylinder is _____.
    a. 4" x 8"
    b. 6" x 12"
    c. 8" x 12"
    d. 4" x 12"
    e. 8" x 16"
    Response _____ Reference _____

6.	When taking a slump test, the slump cone should be filled in _____ equal lifts.
	a.	two
	(b)	three
	c.	four
	d.	five
	e.	none of the above
	Response _____ Reference _____

7.	The most common method of sampling hardened concrete is by _____.
	a.	sampling a broken piece from the structure
	b.	using a Swiss hammer
	c.	using a Windsor probe
	(d.)	extracting cores
	e.	none of the above
	Response _____ Reference _____

8.	Which of the following will aid in reducing water-cement ratio?
	a.	reduce the percent of sand
	b.	use larger-sized coarse aggregate
	c.	use an air-entraining agent
	d.	improve the sand grading
	(e.)	all of the above
	Response _____ Reference _____

9.	When testing compressive strength cylinders in the laboratory, the cylinders should be

	_____.
	(a.)	at room temperature
	b.	between 60 and 80 degrees F
	c.	between 50 and 80 degrees F
	d.	immersed in water
	e.	treated with oil
	Response _____ Reference _____

10.	Compressive strength tests should be made at a location where they will be undisturbed
	for at least _____ hours.
	a.	8
	(b.)	12
	c.	24
	d.	36
	e.	48
	Response _____ Reference _____

11. The thermometer used to obtain the temperature of fresh concrete should be graduated in _____ degree(s) F increments.
    a. $^1/_2$
    b. 1
    c. 2
    d. 3
    e. none of the above
    Response _____ Reference _____

12. How many compressive strength specimens should be made for each test?
    a. one
    b. two
    c. three
    d. four
    e. five
    Response _____ Reference _____

13. Each layer in a slump test should be rodded _____ times.
    a. 10
    b. 15
    c. 20
    d. 25
    e. 30
    Response _____ Reference _____

14. A Swiss hammer has an accuracy of between _____ percent depending on how well it is calibrated.
    a. 2 and 3
    b. 5 and 10
    c. 10 and 20
    d. 5 and 8
    e. 15 and 20
    Response _____ Reference _____

15. Compressive strength test samples for each class of concrete shall be taken not less than once for every _____ cubic yards of concrete.
    a. 50
    b. 100
    c. 125
    d. 150
    e. 200
    Response _____ Reference _____

16.   Maturity methods can be an effective means to determine adequate strength for_____.
        a. form removal
        b. post-tensioning work
        c. sawing joints in slabs-on-grade
        d. controlling accelerated heat curing methods
        e. all of the above
        Response _____ Reference _____

17.   When the lab technician measures and reports 7-day versus 28-day strength data for concrete test cylinders, the technician is reporting_____.
        a. durability data
        b. frequency data
        c. maturity data
        d. petrographic data
        e. none of the above
        Response _____ Reference _____

18.   The 4-inch by 8-inch test cylinder is _____.
        a. easier to cast
        b. requires less sample
        c. easier to handle
        d. requires less field curing space
        e. all of the above
        Response _____ Reference _____

II True/False

19.   In general, the strength of extracted cores is lower than the strength of standard cylinders tested at an identical age.
        T ___✓___   F _____   Reference _____

20.   When sampling fresh concrete, strength tests require at least 2 cubic feet of material.
        T _____   F ___✓___   Reference _____

21.   When sampling fresh concrete for uniformity from a truck mixer, the minimum total sample size should be 4 cubic feet.
        T ___✓___   F _____   Reference _____

22.   When taking core samples of hardened concrete, the diameter of the core must be at least three times the nominal size of the largest aggregate.
        T _____   F ___✓___   Reference _____

23.   Tests do three things—they reveal the quality of a product, they show how uniform the product is and they verify the total volume.
        T _____   F ___✓___   Reference _____

24. A Windsor probe will measure hardness to a greater depth than a Swiss hammer.
    T __✓__ F _____ Reference _____

25. There are two types of air meters in regular use, pressure and volumetric.
    T __✓__ F _____ Reference _____

26. When making specimens for flexural strength of beams, the concrete is placed in two layers with each layer rodded 60 inches.
    T __✓__ F _____ Reference _____

27. The depth of the second layer in a slump test is about 8 inches.
    T _____ F __✓__ Reference _____

28. When using a Swiss hammer, it is usual practice to take 15 readings and average them together.
    T _____ F __✓__ Reference _____

29. The handle of the ball penetration apparatus is 12 inches long.
    T _____ F __✓__ Reference _____

30. An abnormally low unit weight indicates either a high air content or excessive water.
    T __✓__ F _____ Reference _____

31. The size of the specimen for a splitting tensile strength test is the same as that for a compressive strength test.
    T __✓__ F _____ Reference _____

32. When compressive strength cylinders are sent to the laboratory, they are stripped and placed in a 75 degree F dry curing room.
    T _____ F __✓__ Reference _____

33. Most procedures for rapid strength measurement rely on heat to accelerate hydration.
    T __✓__ F _____ Reference _____

34. Maturity methods for determining the strength of concrete are recognized for final evaluation and acceptance of concrete according to the *International Building Code*.
    T _____ F __✓__ Reference _____

35. Maturity methods are recognized as an acceptable method for final evaluation and acceptance of concrete in ACI Standard 318.
    T _____ F __✓__ Reference _____

36. Before maturity of job-placed concrete can be determined, a maturity curve for the specific concrete mix must be developed in the laboratory.
    T __✓__ F _____ Reference _____

37. According to the *International Building Code*, the 4-inch by 8-inch test cylinder is not permitted for final evaluation and acceptance of structural concrete.
T _____   F __✓__   Reference _____

38. A ⁵/₈ inch tamping rod is required for molding the 4-inch by 8-inch test specimens.
T _____   F __✓__   Reference _____

39. The difference in compressive strength between the 4-inch by 8-inch and 6-inch by 12-inch test cylinder size is insignificant.
T __✓__   F _____   Reference _____

40. The ASTM C31 specification requires that test cylinders made in the field shall be 6-inch by 12-inch size, unless otherwise required by the project specifications.
T __✓__   F _____   Reference _____

III Completion

41. Slump is measured in __inches__; a __low__ slump indicates a stiff or dry consistency, a __high__ slump indicates a soft or wet consistency.
Reference _____

42. The ball penetration apparatus should be placed on the _leveled surface_ of the fresh concrete.
Reference _____

43. The slump cone is a metal container that is open at the _____ and the _____ and has a base of _____, a height of _____ and a top of _____.
Reference _____

44. Because important decisions are based on tests results, strict and undeviating _____ of the specific procedures will achieve _____ and _____.
Reference _____

45. A change in coarse aggregate grading may affect the percent of _____ and the rodded _____ of the aggregate, which is reflected in a change in the amount of sand required.
Reference _____

46. Concrete used in an air meter in which water is used to fill the container should not be used for _____ tests or _____ specimens.
Reference _____

47.　When compressive strength specimens are to be consolidated by vibration, the concrete is placed in _____ layers and a vibrating element approximately _____ in diameter is inserted _____ times in each layer for _____ or _____ seconds.
Reference _____

48.　Tests performed on hardened concrete are made in order to _____ or _____ the quality of the hardened concrete.
Reference _____

49.　The basic requirement of any sampling procedure is to obtain a truly _____ sample of the material.
Reference _____

50.　Compressive strength specimen molds should be placed on a _____, _____, firm surface before filling. They should be filled in _____ equal layers, with each layer being rodded _____ times using a _____ diameter steel rod with a _____ tip.
Reference _____

CHAPTER 14
BATCHING AND MIXING THE CONCRETE

Objectives: To understand how materials for concrete are to be handled, the types of batching and control systems in current use, and the types of mixers. Also reviewed are the history, operation and control of ready-mixed concrete, as well as the responsibilities of those involved in all aspects of concrete construction.

Lesson Notes: Size is not a qualifier for the quality mixing of concrete. Quality control may in fact be easier to achieve with smaller batches as opposed to a large operation in which one mistake can result in hundreds of yards of defective concrete.

Key Points:

- When can aggregate that does not meet the standards of ASTM C-33 be used?
- What requirements are covered by ASTM Standard C-33?
- Define "fine aggregate."
- What is the difference between natural and manufactured sand?
- Restate how the maximum size of aggregate is determined.
- What is the purpose of finish screening?
- How might coarse aggregate be contaminated?
- Why should special precautions be taken when taking aggregates from the bottom of a pile?
- Why and for how long should aggregates be allowed to drain?
- How should cement be handled at the plant?
- Of the materials associated with concrete, which almost never presents a problem when storing and handling?
- In what three conditions are admixtures received from manufacturers?
- How should admixtures be stored?
- At what point are superplasticizers usually introduced into a mix?
- How is this requirement affected when using ready-mixed?
- How are pozzolans handled?
- Describe the range of control systems used at a batch plant.
- Of what does a partially automated batching system consist?
- How is this type of system required to be interlocked?
- Of what does a semiautomatic batching system consist?
- How is this type of system required to be interlocked?
- Describe the operation of an automatic batching system.
- How is this type of system required to be interlocked?
- What three devices are found in most modern plants?
- Compare and contrast the differences between these three systems.
- What is the function of a recorder?
- Name the types of recorders.
- How accurate does a recorder have to be?
- What are the Concrete Plant Manufacturers Bureau suggested specifications for recorders?
- In addition to the recording of plant operations, what other information can be obtained from a recorder?
- Describe how the graphic, digital and photographic recorders function.
- Of what does a moisture meter consist?
- How might a moisture meter and a recorder be interconnected?
- Describe the steps in calibrating a moisture meter.
- What is another name for a consistency meter?
- How does a consistency meter work?

- Why is weighing the cement first on a cumulative scale unacceptable?
- Identify the ways in which batching can be done.
- Describe the various ways that an admixture is batched.
- On what should the batch weights of aggregate be based?
- Name the causes of slump variations.
- What is the reason for having a batching sequence?
- Why is it important to check the accuracy of scales and batchers?
- List the steps in checking a scale.
- Before servicing a water-measuring device, what needs to be done?
- How is this calibration accomplished?
- Define "suspense material."
- What should occur after discharge of materials from a batcher?
- What are the three conditions that may exist when transferring batches to a mixer?
- What is a dry batch truck, and how does it work?
- Describe the concerns when using a dry batch truck.
- Name the different types of mixers and how they operate.
- What type of mixer is a paving mixer?
- What type of mixer is a turbine mixer?
- Name the advantages of a turbine mixer.
- State the other names for a horizontal shaft mixer.
- Can a mixer be overloaded?
- On what do uniformity and homogeneity depend?
- List the causes of cement balls.
- Once cement comes in contact with water or damp aggregates, what is the maximum delay in mixing?
- When is more blading required?
- Name the causes of incomplete mixing.
- Why install a timing device on a mixer?
- Define "ready-mixed concrete."
- Of the general factors affecting batching and mixing, which do not apply to ready-mixed concrete?
- List the additional factors that are significant in batching and mixing of ready-mixed concrete.
- For how much volume does ready-mix concrete account?
- Give a brief history of ready-mixed concrete.
- Name the three types of truck mixers.
- Which of these is a nonagitating type?
- With what items should a truck mixer be provided?
- What are the maintenance concerns of a mixer?
- Describe the operation of a mobile batcher.
- What are the advantages of this kind of mixer?
- Define "ribbon loading" and how it works.
- Describe what the sequence should be when adding materials to a mixer.
- During the trip to the job site, what is the speed of the truck mixer?
- Why is overmixing detrimental?
- State the ideal way in which concrete should be discharged from a truck.
- On what is mixing time based for each type of mixer?
- Define what is meant by "agitating speed."
- What is included when considering the total water?
- How are wash and mixing water different?
- Should wash water be allowed as part of mixing water? Why or why not?
- Why is putting water in the spout objectionable?
- Describe how water should be added after a truck has left the batch plant.

- When during or after discharge may water be added to a batch?
- List the information that should be included on a load ticket.
- List the additional information that may be required on the load ticket by the job specifications.
- Investigations have shown certain facts about long-time mixing. What are these facts, and what role does water play?
- Does air content increase or decrease with extended mixing?
- What is the generally agreed-on maximum time frame for mixing?
- How is delayed mixing accomplished?
- Does strict interpretation of the standard for ready-mixed concrete allow for delayed mixing?
- Describe the problems associated with waste disposal and how they are solved.
- Give a brief but thorough description of the producer's and contractor's responsibilities, as well as their joint responsibilities.

CHAPTER 14–QUIZZES

I Multiple Choice

1. Aggregates at the bottom of a pile may be unsuitable because of the intrusion of
    _____.
    a. water
    b. foreign matter
    c. paste
    d. other aggregates
    e. fines
    Response _____ Reference _____

2. A moisture meter usually consists of _____ electrode(s).
    a. one
    b. two
    c. three
    d. four
    e. none of the above
    Response _____ Reference _____

3. Which one of the following is not required on the load ticket?
    a. serial number of the ticket
    b. amount of concrete
    c. MSA
    d. name of the contractor
    e. job name and location
    Response _____ Reference _____

4. The primary function of a recorder is to _____.
    a. check the mix design
    b. make a permanent record of plant operation
    c. verify the quality of materials
    d. indicate the accuracy of the weight and amount of cement
    e. provide quality control
    Response _____ Reference _____

5. One of the producer's responsibilities is to _____.
    a. perform required tests
    b. organize placement and prompt discharge
    c. proportion and batch to meet specifications
    d. provide information on quantity required
    e. all of the above
    Response _____ Reference _____

6.   The use of ready-mixed concrete became widespread after _____.
      a. 1909
      b. 1920
      c. 1930
      d. 1940
      e. 1960
      Response _____ Reference _____

7.   The main advantage of a mobile batch mixer is _____.
      a. accurate mixing
      b. quality control of materials
      c. ease of delivery
      d. portability
      e. all of the above
      Response _____ Reference _____

8.   Fine aggregate is material that passes a No. _____ sieve.
      a.   4
      b.   5
      c.   8
      d.   9
      e. 12
      Response _____ Reference _____

9.   A mixer with a rotating drum that charges, mixes and discharge with its drum axis horizontal is a _____.
      a. plant mixer
      b. vertical shaft mixer
      c. horizontal shaft mixer
      d. tilting mixer
      e. nontilting mixer
      Response _____ Reference _____

10.   The length of time generally agreed on that cement can be exposed to moisture in a mixer is _____.
      a. one hour
      b. one and one-half hours
      c. two hours
      d. two and one-half hours
      e. three hours
      Response _____ Reference _____

II True/False

11.   Aggregates at the bottom of a stockpile located on the ground can be used without concern.
       T _____   F _____   Reference _____

12.   Batch plants that handle more than one type of cement should have each type in a separate compartment.
       T _____   F _____   Reference _____

13.   A ready-mixed concrete producer provides the personnel and equipment to assure continuous production at a rate that meets the needs of the work.
       T _____   F _____   Reference _____

14.   ASTM Standard C 94 provides for mixer performance tests to determine the amount of mixing necessary to completely mix concrete to 70 to 100 revolutions in a stationary mixer.
       T _____   F _____   Reference _____

15.   When used, a superplasticizer must be introduced into the mixer immediately before discharge of the concrete into the receiving equipment.
       T _____   F _____   Reference _____

16.   Trucks used to supply ready-mixed concrete to the job site must be cleaned so that concrete will not accumulate on the shell or around the blades.
       T _____   F _____   Reference _____

17.   Delayed mixing is acceptable under the ACI 318 Standard, provided it is done with close supervision and strict control.
       T _____   F _____   Reference _____

18.   If water is not added, long-time mixing will not affect slump or stiffness.
       T _____   F _____   Reference _____

19.   The suggested mixing time of a 4 cubic-yard mixer is about three minutes.
       T _____   F _____   Reference _____

20.   The ASTM C 94 specification for ready-mixed concrete permits the use of recycled wash water as mixing water in concrete.
       T _____   F _____   Reference _____

III Completion

21. The method of _____ and _____ the cement and aggregates into the _____ has a very important influence on the efficiency of mixing.
Reference _____

22. When batching, cement must be weighed _____; aggregates may be _____, weighing each in turn; and if weighed, water should be weighed on _____.
Reference _____

23. Control systems range from manually controlled individual batchers that depend on the operator's visual observation of a _____ or _____ to fully automated systems that are actuated by a single starting _____ and that stop automatically when the _____ has been reached.
Reference _____

24. Dry-batch trucks should be checked to ensure that their _____ are clean and _____ should be tight to prevent loss of materials and _____ before leaving the plant during rainy weather.
Reference _____

25. Total water in concrete includes free water on the _____, _____ in admixtures, _____ used in hot weather and water added to the batch.
Reference _____

26. The Concrete Plant Manufacturers Bureau described a _____ batching system as a combination of semiautomatic interlocking _____ or of semiautomatic interlocked and _____ batching controls.
Reference _____

27. To promote thorough mixing inside a drum mixer, the _____ should be designed to move the concrete from _____ end of the drum to the _____, with many crossing of _____.
Reference _____

28. There are two potential sources of trouble when aggregate is delivered to the plant by _____: placing the _____ material in a pile, and _____ and _____ being carried into the pile by the truck.
Reference _____

29. A few of the items that are included on a ready-mixed load ticket are the date, _____ number, name of the _____ and the _____, amount of _____, and time _____.
Reference _____

30.   One type of concrete hauling unit has a truck-mounted dump body, which is popularly referred to as a _____ because of the _____ of its corners.
Reference _____

31.   Scales and batching equipment should be kept _____. Binding of _____ or _____ knife edges and _____ causes serious weighing errors.
Reference _____

CHAPTER 15
HANDLING AND PLACING THE CONCRETE

Objectives: To understand the preparation needed prior to placing concrete, the various ways of conveying and pumping, and the proper placement and consolidation of the concrete.

Lesson Notes: When depositing concrete in the forms, the term most commonly used is "pouring"; however, "placing" is the correct term and is more accurate insofar as pouring applies only to a liquid. The use of the word "pouring" originated in the days when wet, sloppy concrete was permitted to flow into place.

Key Points:

- What are the three phases of placing concrete?
- Describe the preparation needed for placing concrete in foundations, especially in regard to free water, soil, and frost and ice.
- How are cast-in-place piles and caissons inspected?
- When may a construction joint be required?
- When a construction joint that is not detailed on the plans is necessary, where is the best location for it?
- The ACI 318 standard provides for installation of construction joints. List these requirements in relation to strength, slabs, beams, girders, columns and walls.
- Why should formed construction joints be avoided?
- Is roughness necessary for a good construction joint? Explain.
- Does reinforcing usually continue through a construction joint?
- What is a shear key? How is it formed?
- Describe the factor that can cause laitance at a construction joint.
- What methods can be used to clean a joint?
- Review Chapter 11 on forms.
- What substance may or may not be used on reinforcing steel?
- When may embedded items be placed in plastic concrete?
- Does this apply to items that are completely embedded?
- What should be covered on final inspection?
- Name the concerns that should be addressed prior to placing concrete.
- What factors must be considered when choosing conveying equipment?
- Which method of conveying minimizes handling?
- Identify the advantages and disadvantages of direct discharge.
- Describe the form of a chute.
- How is conveying affected by the elevation of the chute in relation to the forms?
- Name the most common types of buckets used to convey concrete, how each one works and its advantages.
- Identify the types of belt conveyors.
- What are the chief disadvantages of, or problems with, belt conveyors?
- Describe how a belt conveyor functions.
- The best equipment for moving small amounts of concrete is a wheelbarrow or a buggie. What are their limitations with regard to volume, distance and time?
- Name three other methods of conveying concrete.
- What is one of the chief considerations when placing concrete?
- How should concrete be dropped?
- What are the advantages of concrete pumps?
- What is a small line pump?
- List the capabilities of present day pumps.

- Name the three types of pumps.
- Describe the various types of pump hoses.
- What should be the hose diameter if the MSA is 1 inch? If it is $1^1/_2$ inches?
- When employing a pump, what is the best type of aggregate to use?
- How does aggregate grading affect pumping?
- List the admixtures that improve pumpability.
- Why is oversanding no longer necessary when pumping concrete?
- What can be done to improve pumpability if an angular coarse aggregate is used?
- What is the best slump for pumping concrete?
- What is the most common aggregate size when pumping concrete?
- How does pumping affect slump?
- How should pumping be started?
- What concerns are associated with keeping concrete in a pump hopper?
- What are the causes of line blockage, and how can they be avoided?
- Describe the problems with downhill pumping.
- What is the main problem in pumping lightweight concrete?
- How can this problem be avoided?
- From where does the term "pouring" originate?
- What invention eliminated the need for very wet concrete mix?
- State the basic rule of placing concrete.
- Name the types of equipment used to deposit concrete.
- How quickly should concrete be placed?
- Describe how concrete should be placed in walls, footings or beams of considerable height.
- When might a window be required to place concrete? Describe how this is done.
- How should concrete be placed in deep footings or piles?
- Give a brief description of how best to place monolithic columns and slabs.
- How is concrete placed in large structures such as dams?
- Why should concrete not be placed during a heavy rain?
- What precautions are necessary when placing concrete after rain has started?
- Name the two kinds of vibrators.
- Is vibration required? Justify your answer.
- Describe the characteristics of each of the types of vibrators.
- Describe the characteristics of external vibrators, as well as where they are used and the speed and method of vibration.
- When and where should vibration be applied?
- Against what should a vibrator not be placed?
- How would you handle concrete that has segregated?
- Is overvibration ever a problem? Justify your answer.
- When is concrete revibrated?

CHAPTER 15–QUIZZES

I Multiple Choice

1. The best concrete mix for pumping is a plastic, workable mix with a slump range between _____ inches.
    a. 3 to 6
    b. 4 to 6
    c. 4 to 8
    d. 5 to 7
    e. 2 to 5
    Response _____ Reference _____

2. Chutes should be made of _____.
    a. wood
    b. metal
    c. plastic
    d. aluminum
    e. any of the above
    Response _____ Reference _____

3. One problem associated with belt conveyors is _____.
    a. segregation
    b. consolidation
    c. mortar leakage
    d. motor failure
    e. all of the above
    Response _____ Reference _____

4. Forms should be clean, tight and _____.
    a. wet
    b. staked
    c. properly braced
    d. supported by earth
    e. all of the above
    Response _____ Reference _____

5. Vibrators can be grouped into two classes: _____.
    a. mechanical and electrical
    b. external and internal
    c. pneumatically driven and electrical
    d. pan and screed
    e. table and shaft
    Response _____ Reference _____

6. The most commonly used aggregate in a pump mix is _____ inch(es).
   a. $^3/_4$ or 1
   b. 1 or $1^1/_4$
   c. 1 or $1^1/_2$
   d. $1^1/_2$ or 2
   e. pea gravel
   Response _____ Reference _____

7. High-frequency vibration for consolidation of concrete was introduced around _____.
   a. 1950
   b. 1945
   c. 1940
   d. 1935
   e. 1930
   Response _____ Reference _____

8. Prior to placing concrete when using a pump, the hose should be _____.
   a. primed with water
   b. straight and without radius, bends or kinks
   c. kept at pump level
   d. lubricated with form oil
   e. primed with mortar
   Response _____ Reference _____

9. When using a wheelbarrow to transport concrete, the maximum horizontal distance should be _____ feet.
   a. 100
   b. 150
   c. 175
   d. 200
   e. 250
   Response _____ Reference _____

10. Conveyor belts for placing concrete have an average capacity of about _____ cubic yards per hour.
    a. 20 to 30
    b. 30 to 40
    c. 40 to 50
    d. 50 to 60
    e. 60 to 70
    Response _____ Reference _____

11.   Proper consolidation of concrete decreases_____.
        a. cold joints
        b. honeycombing
        c. entrapped air
        d. segregation
        e. all of the above
        Response_____Reference_____

12.   Concrete is properly vibrated when_____.
        a. concrete surface takes on a sheen
        b. large air bubbles no longer appear at surface
        c. vibrator changes pitch or tone
        d. large aggregate blends into surface
        e. all of the above
        Response_____Reference_____

II True/False

13.   Revibration occurs when the vibrator, in consolidating a layer of concrete, penetrates
        into the layer below to unite the two layers.
        T _____   F _____   Reference _____

14.   When pumping concrete during an extended delay, it is not good practice to run the
        pump every few minutes.
        T _____   F _____   Reference _____

15.   There are two types of piston pumps: hydraulic and mechanical.
        T _____   F _____   Reference _____

16.   When using a bucket to place concrete, the bucket should have a capacity of at least
        one batch.
        T _____   F _____   Reference _____

17.   Prior to placing concrete, excavations for foundations should extend into sound,
        undisturbed soil or rock.
        T _____   F _____   Reference _____

18.   The most common width of a conveyor belt used to place concrete is about 24 inches.
        T _____   F _____   Reference _____

19.   When using wood forms for blockouts, the wood should be clean and dry prior to
        placing the concrete.
        T _____   F _____   Reference _____

20.   If rains starts before concrete placement has been completed, cover the work area with
        tarps until the concrete has set.
        T _____   F _____   Reference _____

21.	Vibrators that are attached to forms and that vibrate the concrete by vibrating the forms are external-type vibrators.
T _____ F _____ Reference _____

22.	Sites that are especially suited for pumping of concrete are those where access is limited or that are crowded with materials.
T _____ F _____ Reference _____

23.	A thin coating of rust on reinforcing steel is detrimental, and dried mortar splashed on the steel must be removed.
T _____ F _____ Reference _____

24.	Hauling buckets on trucks for a considerable distance can cause segregation of the concrete.
T _____ F _____ Reference _____

25.	When vibrating formed concrete, the vibrator should be tilted slightly after contacting bottom of form.
T_____F_____Reference_____

26.	To avoid over vibration, a vibrator should be lifted rapidly from the concrete after each insertion.
T_____F_____Reference_____

27.	Vibration of concrete is acceptable if the vibrator can be easily pushed into the concrete.
T _____F_____Reference_____

III Completion

28.	Roughness is not essential to a good construction joint. A better joint is achieved if the surface of the old concrete is _____ and _____.
Reference _____

29.	Essential to any system of moving concrete from a mixer to forms is to minimize _____, prevent loss of _____ and avoid excessive loss of _____.
Reference _____

30.	In difficult locations, such as on a steep hillside, a pump can easily move the concrete over _____ that would be difficult for a truck to reach.
Reference _____

31.	Cause of line blocks are slump to _____; harsh, unworkable _____; a mix that is too _____; bleeding of the concrete; a long line exposed to the _____; and a long interruption in _____.
Reference _____

32.   A vibrator should not come into contact with the _____ nor held against the
       _____.
       Reference _____

33.   With few exceptions, placing of _____, _____, _____,
       and _____ should be done prior to concrete placement.
       Reference _____

34.   Pumps are currently available with capacities in excess of _____ cubic yards per
       hour, _____ feet vertically and _____ feet horizontally.
       Reference _____

35.   Vibrators should be placed at points that are uniformly _____ close enough
       together to ensure _____ and for _____ seconds
       duration per insertion.
       Reference _____

CHAPTER 16
SLABS ON GROUND

Objectives: To gain an understanding of the requirements for correct placing of concrete on all types of slabs, including suspended slabs.

Key Points:

- What is the most important property of a slab?
- Name the types of ground slabs.
- How must the subgrade be prepared?
- Of what material should the backfill of trenches be?
- What types of soils should be avoided in the subgrade?
- How essential is good drainage to sidewalks, floors and patio slabs?
- How far apart should edge forms be braced?
- At what elevation should the stakes be?
- How are curbs formed?
- Describe the method for setting narrow walkway forms.
- What is a screed?
- What is the difference between a screed and a wet screed?
- What is the slope for interior and exterior slabs requiring drainage?
- Name the materials that can be used for the support of reinforcing in slabs.
- When is a vapor barrier needed?
- What material is normally used as a vapor barrier?
- Describe how the vapor barrier should be installed.
- When would a vapor barrier not be required on a slab?
- When should slabs be air-entrained?
- How much cement and what slump should concrete for exterior walks and driveways have?
- Review each type of floor classification given in Table 16.1. Give the use, special comments and final finish for each class.
- What is the suggested slump for each of these classes?
- Describe the condition of the subgrade prior to placing concrete.
- List the steps in placing concrete in slabs.
- When is concrete ready for final finishing?
- What are a darby, bullfloat, tamper and jitterbug?
- When should a tamper or jitterbug not be used?
- Name the variables for each step of the placing procedure when a superplasticizer is used.
- What is the primary function of joints in slabs?
- Name the three types of joints and their purpose.
- When are construction joints used?
- What is used when a bond across a joint is required?
- Why might this bond be needed?
- What can happen if dowels are not placed perpendicular to the bulkhead?
- When are contraction joints used?
- What is another name for contraction joints?
- Describe four methods for placing contraction joints.
- When a mix has normal shrinkage characteristics, at what distance should contraction joints be placed?
- When are isolation joints used?
- What is another name for an isolation joint?
- How is an isolation joint installed?

- Describe the differences and similarities between these three types of joints.
- What should the specifications for floors contain?
- Name the requirement that all floors have in common.
- Identify the causes of low durability and its results on a slab floor.
- Review the section of Chapter 4 entitled "Agencies of Destruction."
- Define "light-duty floor."
- How is slab thickness determined for medium-duty one-course floors?
- Is reinforcing required in this type of floor? Explain.
- Describe the acceptable ways of installing wire mesh in medium-duty slabs.
- What are the strength and slump requirements for a medium-duty floor?
- Define "two-course heavy-duty floors."
- How is wear resistance obtained for a heavy-duty floor?
- What is expansive soil and how does it react with water?
- Define "suspended slabs."
- What are suspended slabs usually designed to do?
- How do the placing procedures for a suspended slab differ from those for a ground slab?

CHAPTER 16—QUIZZES

I Multiple Choice

1. A concrete floor that is not exposed to heavy loads or to an aggressive environment is a _____ floor.
   a. light-duty
   b. medium-duty
   c. heavy-duty
   d. special-duty
   e. none of the above
   Response _____ Reference _____

2. The maximum recommended slump of a medium-duty floor is _____ inches.
   a. 2
   b. 3
   c. 4
   d. 5
   e. 6
   Response _____ Reference _____

3. The subgrade must be prepared by removing _____.
   a. grass
   b. roots
   c. organic matter
   d. soft soil
   e. all the above
   Response _____ Reference _____

4. A floor slab where industrial vehicular traffic is anticipated should have a _____ finish.
   a. single trowel
   b. float
   c. broom
   d. hard steel trowel
   e. rake
   Response _____ Reference _____

5. Isolation joints allow a slab to _____.
   a. move vertically
   b. move horizontally
   c. move vertically and horizontally
   d. expand
   e. none of the above
   Response _____ Reference _____

6. An interior floor slab should _____ if moisture is present under it.
   a. be built on a vapor barrier
   b. have a 2-inch sand barrier
   c. be built with Type V cement
   d. have adequate subdrains
   e. none of the above
   Response _____ Reference _____

7. When placing concrete, the final compacting following the strike-off is accomplished by the use of a _____.
   a. screed
   b. bullfloat
   c. rake
   d. jitterbug
   e. tamper
   Response _____ Reference _____

8. When installing contraction joints, the groove edges should be __b__.
   a. squared
   b. slightly rounded
   c. tapered
   d. angled
   e. any of the above
   Response _____ Reference _____

9. Where installed in a slab, reinforcing should be supported by __d__.
   a. pieces of stone
   b. metal stakes
   c. wood supports
   d. chairs
   e. any of the above
   Response _____ Reference _____

10. Spacing of contraction joints should not exceed __b__ times the slab thickness where normal shrinkage is anticipated.
    a. 20
    b. 30
    c. 40
    d. 50
    e. 60
    Response _____ Reference _____

II True/False

11. When a new slab is placed adjacent to existing concrete, there must be a separation to allow for movement relative to the old concrete.
    T __✓__   F _____   Reference _____

12. When a slab is water soaked much of the time, a nonpermeable layer should be installed for a depth of 6 inches.
T _____ F ✓ Reference _____

13. Rakes, shovels and hoes are acceptable to spreading concrete.
T _____ F ✓ Reference _____

14. The thickness of a medium-duty one-course floor slab is determined on the basis of the strength and slump of the concrete used.
T _____ F _____ Reference _____

15. After a good floor has been properly cured, its durability cannot be improved by further drying.
T _____ F ✓ Reference _____

16. A floor in a dwelling that is intended to be covered by carpet should be of the same hardness quality as a warehouse floor.
T _____ F _____ Reference _____

17. The primary function of most joints in concrete is to control or minimize cracking and other volume changes, or to permit relative movement of adjacent portions in a structure.
T ✓ F _____ Reference _____

18. The drying shrinkage of the concrete in a large slab will cause random cracks in the slab unless means are provided to relieve this stress.
T ✓ F _____ Reference _____

19. Premolded material in an expansion joint must be at least one-half as wide as the slab is thick and may extend slightly above the slab.
T _____ F ✓ Reference _____

III Completion

20. A suspended slab is one that does not require support by the _____ and must meet the structural requirements of the _____.
Reference _____

21. A wet screed is a strip of concrete about _____ inches wide that is placed just before placing concrete for the slab.
Reference _____

22. Prior to placing a concrete slab, the subgrade should be saturated for ___1 day___ before and _damp_ at the time concrete is to be placed.
Reference _____

23. When floors must be sloped for drainage, interior slabs should have a slope of at least ____ inch per _____, and exterior slabs should have at least _____ inch per _____. Anything less is likely to result in _____.
Reference _____

24. In locations where concrete placing is discontinued, a _____ should be installed and a _____ made. The location of construction joints on a large slab should be _____.
Reference _____

25. The effect of adequate cement on the durability of a floor can be nullified by a lack of _curing_____, high _slump_____, over-vibration or working the surface when _bleed water_____ is present.
Reference _____

CHAPTER 17
FINISHING AND CURING THE CONCRETE

Objectives: To understand the proper application and use of concrete finishing tools and the wear resistance, special treatments and decorative finishes for floors. The materials, time and methods of curing will be reviewed also.

Lesson Notes: Improper curing can ruin what otherwise would be good quality concrete. Unfortunately, it is often neglected or done improperly, thus reducing durability and structural adequacy. Conscientiously following proper curing procedures will result in good durable concrete. Additionally, finishing, if hurried, can turn an attractive product into an unsightly mess.

Key Points:

- Define "finishing."
- What kinds of surfaces are unformed?
- When does the finishing operation begin?
- State the basic law of finishing concrete.
- Do all slabs require edging?
- Where does edging fit in the finishing sequence?
- Describe the purpose of edging.
- Of what materials might an edger be made?
- What is the radius range of an edger?
- How is edging done?
- At what point is grooving begun?
- Of what is a groover made?
- What is the size range of the groove-forming bit?
- How is a guide used when grooving?
- Name the important points of the correct method of grooving.
- What is the third step in finishing?
- Of what materials might a float be made?
- What are the sizes of floats?
- What materials are used for special floats employed for rendered surfaces?
- When should floating start?
- What is the purpose of floating?
- Which material is best for floats?
- Name the last step in finishing concrete.
- What are the sizes of finishing trowels?
- What is the best type of finishing trowel?
- How is the first troweling done?
- How can smoothness of the concrete surface be improved?
- How should bubbles and blisters be treated when troweling?
- How is a nonslip applied?
- Describe the methods, besides brooming or brushing, of applying a nonslip finish to concrete.
- What is the hardness factor of concrete?
- How do the following affect hardness:
    1. No curing at all
    2. Chemical hardening
    3. Application of silica or emery
    4. Application of iron or aluminum oxide
- What is the most common fault of a concrete floor?

- Define "dusting."
- What are the causes of dusting? Describe each in detail.
- How do chemical treatments for dusting work?
- Describe the chemical treatment processes for hardening of a concrete floor that is dusting.
- Give a complete account of the steps to obtaining a heavy-duty floor.
- What is meant by a dry shake coat?
- What is the purpose of a dry shake coat?
- Name the materials and sizes used in dry shake coatings.
- If there are no manufacturer's recommendations, how can a dry shake coating be applied?
- How do liquid hardeners work?
- Should liquid hardeners be considered for any floor slab?
- Name two ways a travertine surface can be obtained.
- Give a brief account of how rock salt can produce a textured surface.
- Describe how simulated flagstone is made.
- Identify the three methods for parting color to concrete.
- Name the materials involved in the dry shake method of coloring concrete.
- Describe how the dry shake method is performed.
- Where is exposed aggregate most effectively used?
- Describe the two methods employed for creating exposed aggregate concrete.
- Briefly describe how to obtain exposed aggregate concrete using the integral and seeding methods.
- Why would a retarder be used in the integral method?
- What should be the MSA in an exposed aggregate slab when the seeding method is used?
- List the six suggestions that apply to any method of exposing aggregate.
- What is terrazzo?
- How much flexibility in color and design is there in a terrazzo finish?
- Where is sand-cushion terrazzo concrete used?
- Describe how a sand-cushion terrazzo concrete floor is installed.
- What are the similarities and differences between a sand-cushion and a bonded terrazzo floor?
- Describe the material and installation needs of a terrazzo topping.
- How do dividers control cracking?
- What can occur if concrete is not properly cured?
- What does curing do?
- Over what period of time does curing occur?
- Name the four methods of curing.
- What condition is water to be in when used to cure concrete?
- What materials can be used for wet coverings?
- List the three types of impermeable sheet materials allowed by ASTM Standard C 171.
- What is the minimum thickness of polyethylene film used for curing concrete?
- Give the various types of sealing compounds listed I ASTM standard C 309.
- Why is continual stirring of sealing compounds required?
- Where might a material be used as both a curing and parting compound?
- When may sodium silicate be used in curing compounds?
- Explain why curing time varies from one job to another.
- What time period is most crucial in concrete curing?
- What are the minimum curing times for various cements?
- What are the two general categories of curing methods? Which method is best?
- Give the positive and negative aspects of the methods of curing that supply added moisture.
- Briefly describe how wet burlap, spray pipes, flooding, wet earth and cotton mats are used to cure concrete.

- Name four common materials used for wet curing with blankets or mats.
- Give the positive and negative aspects of the methods of concrete curing that prevent loss of moisture.
- When is brush application of sealing materials acceptable? Explain.
- When is this type of material applied?
- When can these compounds be thinned?
- Which multi-use material is used in tilt-up construction?
- What is Confilm, and how is it applied?
- What is the usual temperature range of high-temperature curing?
- Identify the concerns when using high-temperature curing.
- What is steam curing?
- After steam curing has been completed, is additional curing beneficial?

CHAPTER 17—QUIZZES

I Multiple Choice

1.   Prior to being subjected to high-temperature curing, concrete should undergo a presetting period after casting of between _____ at normal temperatures.
      a. one to two hours
      b. two to three hours
      c. 24 to 48 hours
      d. 48 to 72 hours
      e. one to two weeks
      Response _____ Reference _____

2.   When exposing aggregate, which of the following should not be done?
      a. using calcium chloride in the concrete
      b. using a surface retarder
      c. testing a sample panel under field conditions
      d. using uniform materials
      e. None of the above
      Response _____ Reference _____

3.   When a heavy-duty topping is required and placement has been delayed, the base slab should be _____.
      a. clean
      b. moist
      c. dry
      d. both a and b
      e. both a and c
      Response _____ Reference _____

4.   Curing methods that prevent loss of moisture entail use of _____.
      a. retarders
      b. insulators
      c. sealing materials
      d. mats and blankets
      e. none of the above
      Response _____ Reference _____

5.   After floating, the next step in the finishing process is _____.
      a. troweling
      b. grooving
      c. edging
      d. brooming
      e. none of the above
      Response _____ Reference _____

6.   The best use of liquid hardeners is on _____.
        a. cured floors
        b. new floors
        c. above grade slabs
        d. older floors
        e. all of the above
        Response _____ Reference _____

7.   After grinding, a standard terrazzo topping should have a minimum thickness of at least _____ inch.
        a. $1\,{}^1/_4$
        b. 1
        c. ${}^3/_4$
        d. ${}^5/_8$
        e. ${}^3/_8$
        Response _____ Reference _____

8.   The normal range of temperatures for high temperature curing is _____ degrees F.
        a. 120 to 160
        b. 100 to 125
        c. 150 to 200
        d. 125 to 170
        e. 175 to 225
        Response _____ Reference _____

9.   _____ solutions are not to be used for curing concrete.
        a. Potassium chloride
        b. Sodium sulfate
        c. Calcium chloride
        d. Sodium silicate
        e. all of the above
        Response _____ Reference _____

10.   All concrete must be _____.
        a. finished
        b. cured
        c. edged
        d. treated
        e. all of the above
        Response _____ Reference _____

11.   Unformed concrete surfaces include _____.
        a. floors
        b. slabs
        c. sidewalks
        d. driveways
        e. all of the above
        Response _____ Reference _____

12. One result of dusting a partly hardened slab with dry cement can be _____.
    a. retarded setting
    b. dry shaking
    c. increased hardness
    d. bubbles
    e. all of the above
    Response _____ Reference _____

II True/False

13. Dusting is caused by weak and soft concrete that results from overfinishing, the use of overly fluid mixes or working the surface while bleed water is present.
    T _____ F _____ Reference _____

14. In a heavy-duty slab, joints in the base slab must be continuous through the wearing course; otherwise the topping will crack.
    T _____ F _____ Reference _____

15. A basic law of finishing concrete is to never use any tools on the fresh concrete while bleed water is present on the surface.
    T ✓ F _____ Reference _____

16. Two of the optimum conditions for high-temperature steam curing are dry steam and a slow temperature rise of not over 60 degrees per hour.
    T _____ F _____ Reference _____

17. Curing compounds are dry mixed when they arrive on the job and should not be agitated after initial mixing.
    T _____ F _____ Reference _____

18. Lean concrete in massive structures requires about four weeks for curing if pozzolans are not used. Normal concrete is best cured for seven days.
    T _____ F _____ Reference _____

19. When exposing aggregate, care must be taken to clean the aggregates without undercutting or loosening them. The maximum exposure is about $1/16$ to $1/4$ inch.
    T _____ F ✓ Reference _____

20. Varnish, lacquers, shellac and surface waxes should not be used on terrazzo.
    T _____ F _____ Reference _____

21. When giving a rock salt finish, the salt is spread on the surface of the concrete at a rate of between 5 and 20 pounds per 100 square feet of area after the slab is finished in the normal manner.
    T _____ F _____ Reference _____

22.   A new trowel is difficult to use until it has been broken in for a few weeks.
       T _____   F _____   Reference _____

23.   It is not unusual to construct a floor that is exposed to especially severe conditions of
       traffic and abrasion in two layers.
       T ___✓___   F _____   Reference _____

24.   Polyethylene film used to cure concrete should consist of two sheets at least 4 mils
       in thickness and be black in color.
       T _____   F _____   Reference _____

25.   Color can be imparted to concrete by paints, stains and pigments incorporated into
       the concrete when it is mixed.
       T ___✓___   F _____   Reference _____

III Completion

26.   _____ produces a radius or rounded edge to the concrete that protects the
       concrete from _____ or other _____.
       Reference _____

27.   The dry shake method of coloring concrete consists of _____ cement,
       _____ and specially graded _____.
       Reference _____

28.   Trowels are made of heat-treated _____ steel or stainless steel and are
       _____ to _____ inches long and _____ to _____ inches wide.
       Reference _____

29.   Curing methods that supply moisture include _____, _____ and
       other moisture-retaining _____.
       Reference _____

30.   A dry shake or dust coat can be applied to a one-course slab to give it a high
       resistance to _____ and _____. Application of a dry shake is
       spread on the floated slab _____ the bleed water has _____.
       Reference _____

31.   Materials that can be used for curing concrete include _____,
       _____ compounds, and various _____ and _____.
       Reference _____

32.   A grooving tool is usually made of _____, _____ or _____,
       and is usually _____ inches long with ends _____ slightly to facilitate its
       use.
       Reference _____

33. Aggregate for heavy-duty floors must be _____ and _____, consisting of _____, _____ or similar natural rock particles, or a manufactured product.
Reference _____

34. When using sealing compounds to cure concrete, the compounds should be of a consistency suitable for _____, should be relatively _____, should adhere to a vertical or horizontal _____ concrete surface, and should not react _____ with the concrete.
Reference _____

35. Moist curing after steaming improves _____ and _____, and should be utilized if possible. The greatest advantage of steaming occurs during the _____ and soon reaches a point of diminishing returns.
Reference _____

CHAPTER 18
THE STEEL REINFORCEMENT

Objectives: To give a general overview of the kinds of reinforcing used, how it is fabricated, and its placing, handling and inspection. Also, to provide a brief look at fiber, galvanized and epoxy-coated reinforcement.

Lesson Notes: Perhaps the most important aspect of placement of reinforcement is that it must be installed exactly per the approved plans and engineering details. Substitution of sizes, cutting, bending, splicing and relocation should never be allowed unless approved by the engineer and the building official.

Key Points:

- Why use reinforcement in concrete?
- At what location in a beam is reinforcement usually placed?
- What other purposes, besides tensile stresses, might reinforcing serve?
- What is light reinforcing used to distribute cracks and make them smaller called?
- Of what configuration are stirrups, and how are they placed?
- How are columns reinforced?
- Name and describe the two types of reinforcing bars.
- To which standards are deformations on bars to conform?
- What are the nominal diameters of #4, #6 and #9 bars?
- What does the grade of steel indicate?
- What is the minimum yield strength of a grade 60 bar?
- Define "yield point" and "ultimate tensile strength."
- Name the different grades of reinforcement.
- Which grade(s) of reinforcement do not need a grade mark on the steel?
- Review Figure 18-4, and identify what each of the marks on a piece of reinforcement mean.
- Define "WWR," and explain its features.
- Describe what each of the numbers and letters mean in a piece of WWR 6 x 12-W16 x W26.
- What is the substitute letter for deformed wire?
- Name the advantages of WWR.
- What is a bar mat?
- What attributes must a bar support possess?
- Of what materials may bar supports consist?
- Review Figure 18-15 and list the different types of supports for WWR.
- What is a sand plate?
- What are the three classes of metal bar supports?
- Compare and contrast the three classes of bar supports.
- How is reinforcing kept from moving after placement?
- What are the most common sizes of tie wire?
- What is the normal length of reinforcement supplied by a manufacturer?
- Define "placing drawings."
- What is contained in the placing drawings?
- Where is steel fabricated?
- Where is wall steel usually detailed?
- What is a bar list?
- From what is this list derived, and what does it contain?
- Of what does a reinforcing schedule consist?
- What does fabrication begin?
- How is steel bent?

- Read ACI 318 Sections 7.1 and 7.2 regarding standard hooks and minimum bend diameters. Detail the requirements for hooks and bends.
- What could occur if steel is not cut and bent accurately?
- Review the tolerances chart on *Concrete Manual* page 238.
- What are bundled bars?
- What information should the tag on bundled bars contain?
- What is the first action that should be taken by the inspector?
- What is manifest?
- What is contained in a manifest?
- How is reinforcing stored?
- In what order should steel be stored?
- Of the following list, which item(s) is acceptable on reinforcing? Oil, grease, light rust, paint, mill scale.
- When should a bar be rejected for use?
- When may reinforcing be heated for bending?
- After heating, how should a bar be cooled?
- Which types of tests should be done on reinforcing?
- What should be inspected and verified on each shipment of reinforcing?
- How should reinforcing layers be separated?
- What types of materials should not be allowed to hold reinforcing in position?
- When is welding of crossing bars allowed?
- At what point should inspection of the steel begin?
- Give details of the (two) incorrect and correct methods of placing steel in slabs.
- When is field bending of partially embedded reinforcing acceptable?
- What is mill scale?
- What is the maximum amount of rust that is acceptable on steel?
- How is steel cleaned of grease or of oil?
- Define "cages."
- Describe the code requirements for fixing steel in forms for slabs, walls and footings.
- How is the steel assembled in a two-curtain wall?
- Define "mat."
- How are dowels held in place?
- Define "splicing."
- Name the three general types of splices, and state the differences between them.
- How are splices in adjacent bars done?
- What is the most common type of splice?
- What criteria are followed when using a mechanical splice?
- What are the two types of welded splices?
- Describe a potential problem of welded splices.
- May crossing bars be welded?
- How is steel to be tied together?
- How is distance to forms or subgrade maintained?
- What is meant by the term "dobies"?
- On what does the selection of supports and spacers depend?
- List some of the acceptable ways to support reinforcing.
- Define "tie."
- What is the purpose of tying reinforcing?
- Do all intersections have to be tied?
- Describe the methods of tying steel.
- Why is placing the steel within code tolerances important?

- If the approved plans required a concrete cover over steel to be three inches from the bottom of a footing, give the dimensions within which the steel must be placed. Refer to the tolerance table on *Concrete Manual* page 243.
- What are the usual tolerances for stirrups, column ties and steel in flat slabs?
- What is the purpose of providing concrete cover over reinforcing?
- Under what conditions should cover be increased?
- Where is wire fabric usually used?
- How is wire fabric installed in slabs?
- Describe how wire fabric is placed when used as temperature steel.
- How is wire fabric lapped?
- What is the most common use of heavy fabric?
- Give the various ways in which wire fabric is used.
- Specify the correct and incorrect placement procedures for wire fabric.
- Name the three types of fiber reinforcement in use.
- Which of these is the one most commonly used?
- Why use galvanized reinforcing?
- Are there any concerns with using galvanized steel?
- For what use was epoxy-coated reinforcing first developed?
- Name some of the structures where epoxy-coated steel is used.
- Describe the special precautions necessary when using epoxy-coated steel.

CHAPTER 18—QUIZZES

I Multiple Choice

1. Excessive rusting of the reinforcement weakens the steel and also causes _____ that may result in spalling and cracking.
    a. small voids where
    b. an expansion in volume
    c. a loss of water proofing
    d. a loss of durability
    e. all of the above
    Response _____ Reference _____

2. Which of the following does not interfere with steel bonding to concrete?
    a. paint
    b. grease
    c. mill scale
    d. oil
    e. rust
    Response _____ Reference _____

3. Epoxy-coated reinforcement should be checked for _____.
    a. proper mechanical splices
    b. rust
    c. smoothness
    d. damaged coating
    e. all of the above
    Response _____ Reference _____

4. A #5 bar has an approximate diameter of _____ inch and a #8 bar one of _____ inches.
    a. $5/8$, 1
    b. $5/16$, $1/2$
    c. $1/2$, $3/4$
    d. $3/4$, $1^1/4$
    e. none of the above
    Response _____ Reference _____

5. An advantage of using WWR is _____.
    a. lighter weight
    b. ease of use in columns and beams
    c. increased tensile strength
    d. speed and ease of installation
    e. ease of use in transverse structures
    Response _____ Reference _____

6. Welded splices can be either lap welds or _____ welds.
    a. proprietary
    b. mechanical
    c. tied
    d. hooked
    e. butt
    Response _____ Reference _____

7. To resist movement or displacement, reinforcing bars must be _____.
    a. supported
    b. welded together
    c. tied together
    d. hooked
    e. any of the above
    Response _____ Reference _____

8. Field bending is apt to result in _____.
    a. loss of ductility
    b. loss in compressive strength
    c. loss of bond
    d. increased lap slices
    e. expansion
    Response _____ Reference _____

9. Billet steel is available in Grades_____.
    a. 35, 40 and 50
    b. 40, 50 and 60
    c. 40, 60 and 75
    d. 40, 60 and 80
    e. 60, 75 and 90
    Response _____ Reference _____

10. In addition to the two main ribs, a reinforcing bar may have a third rib. This indicates _____.
    a. type of steel
    b. Grade 60
    c. Grade 75
    d. rail steel
    e. low-alloy steel
    Response _____ Reference _____

11. Factory-made wire bar supports may be made of _____.
    a. plain wire
    b. galvanized wire
    c. stainless steel wire
    d. all of the above
    Response _____ Reference _____

12. A reinforcing bar shipment from a fabricator will be accompanied by a list known as
    a _____.
    a. manifest
    b. invoice
    c. trip ticket
    d. delivery ticket
    e. none of the above
    Response _____ Reference _____

13. The most widely used reinforcing bars are_____.
    a. axle-steel
    b. billet-steel
    c. carbon steel
    d. low-alloy steel
    e. rail steel
    Response _____ Reference _____

14. A reinforcing bar marked **S&W** conforms to ASTM specification_____.
    a. A 615
    b. A 616 and A 617
    c. A 706
    d. A 615 and A 706
    e. A 996
    Response _____ Reference _____

15. A #22 matric reinforcing bar is the same size as a_____inch-pound reinforcing
    bar.
    a. #5
    b. #6
    c. #7
    d. #8
    e. none of the above
    Response _____ Reference _____

16. The equivalent metric grade mark for the inch-pound grade mark 75 is_____.
    a. 3
    b. 4
    c. 5
    d. 42
    e. 52
    Response _____ Reference _____

17. According to the placing drawing on page 236 of the *Concrete Manual*, the required stirrups for grade beam GB1 are indicated as_____.
a. #3@5"
b. #3@6"
c. #4@5"
d. #4@6"
e. none of the above
Response _____ Reference _____

18. According to the placing drawing on page 236 of the *Concrete Manual*, the required reinforcing (each way) for footing F2 is indicated as_____.
a. 12#6
b. 24#6
c. 10#7
d. 20#7
e. 16#7
Response _____ Reference _____

19. According to the placing drawing on page 236 of the *Concrete Manual*, the footing dowels for column D1 may extend_____vertically into the column and be within acceptable tolerance.
a. 2'-5"
b. 2'-6"
c. 2'-7"
d. 2'-8"
e. 2-9"
Response _____ Reference _____

20. Suggested minimum spacing of supports for D9 WWR @14" wire spacing used in slab-on-ground applications is_____.
a. 2 to 3 ft
b. 3 to 4 ft
c. 4 to 6 ft
d. 6 to 8 ft
e. none of the above
Response _____ Reference _____

21. For information on stainless steel welded wire reinforcement, the inspector should refer to ASTM_____.
a. A 185
b. A 497
c. A 955
d. A 1022
e. none of the above
Response _____ Reference _____

22. If a mill test report is not available, welding of #6 billet steel rebars is permitted if the bars are preheated to _____°F.
    a. 100
    b. 200
    c. 300
    d. 400
    e. 500
    Response _____ Reference _____

23. If a mill test report is not available, welding of #6 low-alloy steel rebars is permitted if the bars are preheated to _____°F.
    a. 50
    b. 200
    c. 300
    d. 500
    e. no preheat required
    Response _____ Reference _____

24. If the design drawings for an 8-inch concrete tilt-up panel indicate a $1^1/_2$ inch cover to the vertical rebars, the minimum acceptable measured cover is_____ inch.
    a. $^3/_4$
    b. 1
    c. $1\text{-}^1/_8$
    d. $1\text{-}^1/_4$
    e. none of the above
    Response _____ Reference _____

25. If the design drawings for a 24-inch deep spandrel beam at the perimeter of an elevated slab indicate a clear cover of $1^1/_2$ inch to the bottom reinforcing bars, the minimum acceptable measured cover is_____ inches.
    a. 1
    b. $1\text{-}^1/_8$
    c. $1\text{-}^1/_4$
    d. $1\text{-}^3/_8$
    e. none of the above
    Response _____ Reference _____

26. If the design drawings for a structural slab indicate that the bottom bars of the end span are to be located 3'-0" from the center of the interior column support, the minimum acceptable measured distance is_____.
    a. 2'-8"
    b. 2'-9"
    c. 2'-10"
    d. 2'-11"
    e. 3'-0"
    Response _____ Reference _____

27.   According to ACI Standard 117, if the design drawings indicate a lapped splice length of 24 inches for two #5 bars, the minimum acceptable measured lap length is _____ inches.
        a. 20
        b. 21
        c. 22
        d. 23
        e. none of the above
        Response _____ Reference _____

II True/False

28.   Bar mats are similar to WWR except for the fact that, being made of reinforcing bars, they are heavier.
        T _____   F _____   Reference _____

29.   All bends are made with the steel at normal room temperature except in cold weather, when hot bending is permitted.
        T _____   F _____   Reference _____

30.   Tying the steel is done after it has been placed and spaced properly.
        T _____   F _____   Reference _____

31.   The clear distance between a reinforcing bar and the surface of the concrete is the cover. The purpose is primarily to protect the steel from weathering.
        T _____   F _____   Reference _____

32.   Heavy wire fabric comes in flat sheets and is used extensively in pavements.
        T _____   F _____   Reference _____

33.   A light coating of rust can decrease bond as well as cause spalling and cracking.
        T _____   F _____   Reference _____

34.   Welded-wire fabric is identified by denoting smooth wire with the letter "F," followed by a number indicating the cross-sectional area in hundredths of a square inch.
        T _____   F _____   Reference _____

35.   Class 1 metal supports are plastic protected steel wire bar supports intended for use in severe exposures or in situations requiring sandblasting of the concrete surface.
        T _____   F _____   Reference _____

36.   Bar supports for epoxy-coated reinforcing bars should be coated with a dielectric material such as plastic.
        T _____   F _____   Reference _____

37. A standard hook can be a 180-degree bend plus 4_{db}, but not less than a $2^1/_2$ inch extension at the free end of the bar.
T _____ F _____ Reference _____

38. It is sometimes advantageous to assemble the steel into caissons in which the bars, stirrups and other elements can be tied together at a convenient assembly location.
T _____ F _____ Reference _____

39. If a reinforcing bar appears to have rusted excessively, a sample should be cleaned and weighted to determine compliance with the reference specifications.
T _____ F _____ Reference _____

40. Reinforcing bars are cold rolled into bar size and deformations.
T _____ F _____ Reference _____

41. Older specifications for rail steel (A616) and axle steel (A617) have now been combined in new ASTM A996.
T _____ F _____ Reference _____

42. USA-produced metric reinforcing bars are approximations of the inch-pound bar diameter in meters (m).
T _____ F _____ Reference _____

43. If the structural drawings indicate a #9 reinforcing bar and the iron worker is placing a bar marked 19, the inspector should notify the contractor of the incorrect bar size.
T _____ F _____ Reference _____

44. If the structural drawings indicate a Grade 75, #14 reinforcing bar and the iron worker is placing a bar marked 43 with a grade mark 5, the inspector should notify the contractor of the incorrect bar.
T _____ F _____ Reference _____

45. USA-produced reinforcing bars furnished on the construction project most likely will be soft metric.
T _____ F _____ Reference _____

46. Epoxy coating of reinforcement is an acceptable surface condition of reinforcement.
T _____ F _____ Reference _____

47. According to the AWS welding code for reinforcing steel, #18 rebars are never permitted to be field welded unless the bars are preheated and the preheat temperature has been determined based on the chemical composition (Carbon Equivalent) of the bars.
T _____ F _____ Reference _____

48. ACI Standard 117 is a legally adopted standard of the *International Building Code*.
T _____ F _____ Reference _____

49.   According to ACI Standard 117, the spacing between individual bars of a group of bars to be uniformly spaced at 14 inches may vary between 11 inches and 17 inches.

T _____   F _____   Reference _____

50.   Fiberglass reinforcement is permitted in ACI Standard 318 for structural concrete.

T _____   F _____   Reference _____

51.   FRP rebar significantly improves the longevity of concrete structures where corrosion is a major factor.

T _____   F _____   Reference _____

III Completion

52.   Field bending of bars partially embedded in concrete is _____ by code, except as shown on the approved plans or as permitted by the _____ with the concurrence of the _____.

Reference _____

53.   The minimum yield designation for Grade 60 reinforcing can be marked on the bar by either _____ longitudinal line or the number _____. Grade 75 can be marked by either _____ longitudinal lines or the number _____.

Reference _____

54.   When wire fabric is supplied to the job in rolls, it is rolled out, then draped from a position near the top of the slab over the _____ to the bottom of the slab at _____, keeping the required _____ at each location.

Reference _____

55.   The use of epoxy-coated reinforcing was developed for use in highway bridge decks where concrete is subjected to severe exposures from _____, _____ and _____.

Reference _____

56.   Steel should be stored on _____ or other _____ off the ground to protect it from _____ and _____ on the jobsite and in locations where it may be splattered with _____. Long storage periods will result in excessive _____ or contamination.

Reference _____

57.   A bar list is a bill of materials or a list of _____ covering a portion of the structure. Bars are classified as to _____, _____, and whether they are _____ or _____.

Reference _____

58. Although accuracy is essential, it is necessary to allow for slight inaccuracies in the
_____. These allowances are called _____. The typical
tolerance for a straight bar is plus or minus _____ inch.
Reference _____

59. Reinforcing steel must be secured in place. Distances from subgrade and forms
should be maintained by the use of _____, _____, _____ or
other approved _____.
Reference _____

60. Heating in order to bend reinforcing can only be done when approved by the
_____ with the concurrence of the _____. If heating is
approved, bars should be heated _____ and air cooled _____.
Reference _____

61. Reinforcement is used to control cracks in slabs caused by _____ and
_____ of the concrete resulting from temperature _____. The
reinforcement does not prevent _____.
Reference _____

62. Grades of reinforcing steel are specified by the_____and must be indicated on
the_____and_____.
Reference_____

CHAPTER 19
HOT AND COLD WEATHER CONCRETING

Objectives: To obtain an understanding of the requirements for placing concrete in hot and cold weather, and how to minimize the effects of—and how to control and protect concrete in—weather extremes.

Lesson Notes: It is best to delay placing concrete when weather extremes occur; however, if placement must proceed, it takes a little extra effort to obtain good durable concrete.

Key Points:

- When extremes in weather occur, should concrete placement be suspended?
- Read ACI 318 Sections 5.12 and 5.13.
- What is considered hot weather for placing concrete?
- What is the object of taking precautions to protect concrete at high temperatures?
- List the possible undesirable effects of hot weather on concrete.
- Does hot weather affect strength? Explain.
- How much additional mixing water might be required for a temperature increase of 10 degrees F?
- What will be the result of adding this water to the concrete?
- Explain how cracking and shrinkage is aggravated during hot weather.
- Will hot weather affect concrete after it has hardened?
- How is trouble-free concrete achieved during hot weather?
- Where does control of the temperature of concrete start?
- Describe the ways in which controlling the aggregate temperature can be a benefit.
- How is mixing water kept cool?
- May ice ever be used to cool concrete?
- Which type of admixtures are used to best advantage during hot weather?
- How is heat increased during mixing and delivery?
- Explain the steps that can be taken to reduce heat during these phases.
- List the things that must be planned for placing and finishing concrete in hot weather.
- How do fog nozzles help protect concrete from the effects of hot weather?
- What is the best curing during hot weather?
- Review the summary of hot weather precautions given on *Concrete Manual* page 252.
- At what temperature does cold weather become a concern?
- How does cold weather affect the hydration process?
- How is strength affected by cold weather?
- During what period of time should concrete be protected from cold weather?
- Describe how cold concrete can be beneficial?
- List the indirect effects of cold weather on the durability of concrete.
- What is the best means of heating concrete when freezing temperatures are expected?
- How are aggregates heated?
- When should preparation for cold weather begin?
- What should be the minimum temperatures for concrete placed in thick and thin members?
- What type of cement is most frequently used to reduce the time protection from cold weather?
- When should calcium chloride not be used to accelerate setting time?
- Is air-entrainment desirable for cold weather concrete?
- What admixture is used to lower the freezing temperature of concrete?
- How would a frozen subgrade affect concrete?
- Name some common materials used to protect concrete from freezing.
- List the best means of providing heat in a protective enclosure. What is the worst?
- How long must minimum temperatures be maintained?

- Name the materials used to insulate concrete.
- How are these materials used?
- Should forms be left in place during cold weather?
- Is water curing desirable?
- Review the summary of cold weather precautions listed on *Concrete Manual* page 254.

CHAPTER 19—QUIZZES

I Multiple Choice

1.   Which of the following is not an effect of hot weather?
        a. accelerated setting
        b. increased plastic shrinkage
        c. lower volume of mixing water
        d. rapid slump loss
        e. reduced strength
        Response _____ Reference _____

2.   Considerations for cold weather should begin when the temperature drops below
            _____ degrees F.
        a. 25
        b. 32
        c. 40
        d. 45
        e. 50
        Response _____ Reference _____

3.   Concrete should never be placed _____.
        a. on unreinforced slabs
        b. on a frozen subgrade
        c. during hot weather over 95 degrees F
        d. during cold weather below 25 degrees F
        e. all of the above
        Response _____ Reference _____

4.   Which of the following should not be used to accelerate setting of a prestressed
        concrete member in cold weather?
        a. air-entrainment
        b. calcium chloride
        c. water-reducing admixture
        d. steam
        e. curing compounds
        Response _____ Reference _____

5.   The best way to minimize the effects of hot weather is to cool the _____.
        a. mixing water
        b. sand
        c. coarse aggregate
        d. cement
        e. subgrade
        Response _____ Reference _____

6. If the temperature of a 10 cu yd batch of fresh concrete (in transit from the batch plant to the job site) increases from 50 degrees F to 75 degrees F, an additional_____gallons of water will be required to maintain the same slump. Water weighs 8.33 lb per gallon.
   a. 10
   b. 20
   c. 30
   d. 40
   e. none of the above
   Response _____ Reference _____

7. In the absence of special precautions, undesirable cold weather effects may include_____.
   a. slower setting
   b. slower strength gain
   c. permanent damage due to early freezing
   d. reduced durability
   e. all of the above
   Response _____ Reference _____

II True/False

8. Concrete needs about 7 pounds more water for each 10 degree F rise in temperature.
   T _____ F _____ Reference _____

9. When heating mixing water, the temperature of the water should exceed 175 degrees F.
   T _____ F _____ Reference _____

10. During hot weather, plans must be made so that concrete can be received and placed as rapidly as possible. All equipment should be of adequate capacity, and a sufficient number of workers of all necessary classes should be on hand.
   T _____ F _____ Reference _____

11. High temperature can adversely affect the strength, durability and cracking of concrete, and its ultimate strength may not be as high as that of concrete placed at moderate temperatures.
   T _____ F _____ Reference _____

12. When curing concrete during hot weather, allowing the surface to dry between applications of water is not detrimental to the concrete except when Type III cement is used.
   T _____ F _____ Reference _____

13. If concrete mix proportions for a specified strength and slump were determined at a laboratory temperature of 50 degrees F, and the actual temperature at time of batching is 75 degrees F, additional cement and water will be required to maintain the specified strength and slump.
T _____   F _____   Reference _____

III Completion

14. Because uniform heating of aggregates is too difficult, heating of the aggregates _____ be done when heating of the _____ alone would ensure delivery of the concrete at the required temperature.
Reference _____

15. The results of observations show that concrete made and cured at temperatures of between _____ degrees F has a later higher strength than that of _____ cured concrete.
Reference _____

16. Especially during hot weather, the amount of mixing of concrete should be the minimum that can achieve the necessary _____ and _____, and _____ must be avoided.
Reference _____

17. Inadequate precautions during hot weather can have an appreciably _____ effect on durability, the resistance to freezing and thawing cycles, and a _____ resistance to attach by _____ solutions.
Reference _____

18. The indirect effects of low temperatures include cracking of dehydrated areas caused by a lack of _____ of the surface from heaters and freezing of corners and edges of green concrete that has _____ but is still saturated with water and has _____.
Reference _____

CHAPTER 20
PRECAST AND PRESTRESSED CONCRETE

Objectives: To obtain an understanding of the types of precast concrete units, the purpose of placing drawings, and the forms, fabrication and curing of precast units. Also discussed will be prestressed concrete, including the forms, molds, handling and erection of pretensioned concrete and posttensioned concrete.

Lesson Notes: For more details on the installation of unbonded post-tensioned tendons the reader is referred to the *Field Procedure Manual for Unbonded Single Strand Tendons.* Also, much additional information and explanation can be obtained by studying the *Posttensioning Manual* and the *Manual for Quality Control for Plants and Production of Precast and Prestessed Concrete Products*. Refer to the reference section at the back of the *Concrete Manual* for the relevant addresses.

Key Points:
- Study ACI 318 Chapter 18.
- Define "precast concrete."
- What types of elements are included in this definition?
- What is the most common method of producing precast prestressed concrete?
- What has led to the growth of the precast concrete industry?
- Name some of the reasons for the increased desirability of precast concrete.
- List the advantages of prestressed concrete when compared with conventional concrete.
- Describe the ways that precast units can be classified.
- What is the difference between load-bearing and nonload-bearing members?
- List the types of structural, architectural, framing and miscellaneous precast units.
- For what are tees used?
- What composes a window wall, and how is it attached?
- Where and for what are screen walls used?
- List the items that are included in the general category of precast concrete.
- From what type of concrete can precast units be made?
- Explain the reason for using placing drawings.
- Who should review placing drawings?
- State the reasons for reviewing placing drawings.
- Describe all the items that placing drawings should contain.
- Of what might the formwork for precast units consist?
- On what does the selection of form materials depend?
- Give the advantages and disadvantages for using wood, fiberglass and steel.
- When required, what type of internal ties should be used?
- What concerns are associated with form oils for precast concrete?
- Give the following: most commonly used cement, types of aggregates, types of admixtures, MSA, average slump and range of psi.
- How are materials batched?
- Where is reinforcing for precast items usually assembled?
- Give a brief description of the Schokbeton process.
- How do extruding machines work?
- List the aspects to be covered by the inspector at final inspection prior to placing concrete for precast members.
- What is the most frequently used method of curing precast elements?
- What is prestressed concrete?
- Compare and contrast pretensioning and posttensioning.
- Identify the three most common types of prestressing steel and the standard that covers them.
- Which of these is the most widely used for buildings?

- What are the most commonly used grades of each type of prestressing steel?
- Define "elastic modulus."
- What is the average elastic modulus of prestressing steel?
- What should the frequency of reports be for prestressing strand, wire and bar?
- How should prestressing steel be protected?
- Describe what a casting bed is, of what it consists and why it is utilized.
- Describe the use of bulkheads in casting beds.
- How is prestressing steel elongated?
- How is the amount of elongation determined?
- What may be a source of error in measuring jacking forces?
- What is detensioning?
- What is the difference between multiple- and single-strand detensioning?
- To minimize cracking, what is important in developing a detensioning pattern?
- When are temperature variations around casting beds of importance?
- What is the acceptable amount of broken wires or strands in prestressing steel?
- What information should be included on precast units before being placed in storage?
- Give the basic requirements for handling prestressed units at the plant.
- Describe how units should be stored and how loading on transports should be accomplished.
- What is the first concern after a unit is placed in a structure?
- List the qualities of all prestressed connections to a structure.
- When space is available, precasting can be done on site. What are the advantages to this practice?
- What are the concerns when using precast forms and molds?
- Name the two types of posttensioning.
- What is the difference between them?
- What are the most common sizes of unbonded single-strand tendons?
- Describe of what a tendon consists.
- What is the rubber or plastic block used?
- How is the steel in an unbonded system protected?
- What information do the shop drawings for unbonded tendons contain?
- Describe in detail how unbonded tendons are shipped, labeled and placed.
- What are the concerns when welding near unbonded tendons?
- How is concrete placed in an unbonded tension slab?
- What admixture(s) should not be used in concrete placed in an unbonded slab system?
- When is shoring removed after placing concrete?
- Give a brief description of the stressing operation.
- What are the distinct construction phases in a posttensioning system?
- List the guidelines for each phase.

CHAPTER 20—QUIZZES

I Multiple Choice

1. Which of the following must not be used in posttensioned concrete?
    a. air-entrainment
    (b.) calcium chloride
    c. Type III cement
    d. graded aggregates
    e. all of the above
    Response _____ Reference _____

2. Sheathing of unbonded prestressing tendons must prevent _____ during concrete placement.
    a. spalling
    (b.) intrusion of cement paste
    c. fracturing of anchorages
    d. displacement of tendons
    e. stressing of tendons
    Response _____ Reference _____

3. Preassembled posttensioning tendons are usually shipped to the site in _____ foot diameter coils.
    a. 3
    b. 4
    c. (5)
    d. 6
    e. 7
    Response _____ Reference _____

4. For pretensioning tendons, there must not be a difference of more than _____ percent between stress computed from jack pressure and stress computed from measurement of elongation.
    a. five
    b. six
    (c.) seven
    d. eight
    e. nine
    Response _____ Reference _____

5. The minimum concrete cover between a prestressing tendon and on opening in a slab should not be more than _____ inches.
   a. 4
   b. 5
   c. 6
   d. 7
   e. 8
   Response _____ Reference _____

6. Prestressing force must be determined by _____.
   a. visual observation of tendon stress
   b. measurement of tendon elongation
   c. use of a calibrated dynamometer
   d. both a and b
   e. both b and c
   Response _____ Reference _____

7. Sheathing of unbonded tendons in prestressed concrete must _____.
   a. be within 2 inches of each end
   b. be within 12 inches of each end
   c. not be allowed
   d. be with duct tape if within 12 inches of an end
   e. be over the entire length
   Response _____ Reference _____

8. Any difference between tendon elongation and jacking force on a calibrated gage must not exceed _____ percent for posttensioned concrete.
   a. two
   b. four
   c. five
   d. seven
   e. ten
   Response _____ Reference _____

9. Unbonded prestressing tendons must be coated with _____.
   a. a light oxide
   b. cement paste
   c. material to ensure corrosion protection
   d. paint
   e. galvanizing
   Response _____ Reference _____

10. Which one of the following is not considered to be a precast structural unit?
    a. mullion
    b. box unit
    c. stemmed unit
    d. girder
    e. joist
    Response _____ Reference _____

11. The most common method of curing in precasting plants is _____.
    a. mechanical
    b. high-temperature
    c. chemical
    d. moist
    e. all of the above
    Response _____ Reference _____

12. Immediately after placement of precast units in a structure, _____ must be accomplished.
    a. grouting
    b. temporary bracing
    c. final welding
    d. permanent bracing
    e. none of the above
    Response _____ Reference _____

13. Type _____ cement is most commonly used for prestressed concrete.
    a. II
    b. III
    c. IV
    d. V
    e. none of the above
    Response _____ Reference _____

14. Construction details for posttensioned building elements prepared by the posttensioning subcontractor are referred to as_____.
    a. installation drawings
    b. placing drawings
    c. shop drawings
    d. structural drawings
    e. none of the above
    Response _____ Reference _____

15. For the installation drawing shown on page 270 of the *Concrete Manual*, indicated strand stressing number (43) consists of_____strands.
a. one
b. two
c. three
d. four
e. five
Response _____ Reference _____

16. If the installation drawings for a posttensioned slab indicate an elongation $\Delta = 6\text{-}1/2$ inches for an 81 foot tendon, the range of acceptable measured elongations (minimum – maximum) is _____.
a. 6"-7"
b. $6\text{-}1/8" - 6\text{-}7/8"$
c. $6\text{-}1/4" - 6\text{-}3/4"$
d. $6\text{-}3/8" - 6\text{-}5/8"$
e. none of the above
Response _____ Reference _____

II True/False

17. Differences in the modulus of elasticity of different production lots of steel is a source of error in measuring jacking forces.
T _____ F _____ Reference _____

18. Prestressed concrete requires less reinforcing steel and concrete to produce units with strength equal to conventionally reinforced concrete.
(T) _____ F _____ Reference _____

19. At least two certified test reports should be furnished for each 20-ton production of each size of prestressing steel.
T _____ (F) _____ Reference _____

20. In general, most prestressing strands are tensioned to about 70 percent of ultimate strength.
(T) _____ F _____ Reference _____

21. Conduits and other utilities cannot be accommodated in precast concrete.
T _____ (F) _____ Reference _____

22. Positioning of the prestressing strands is not critical in precast units.
T _____ (F) _____ Reference _____

23. Unbonded single-strand tendons used in posttensioned slabs usually consist of $3/8$-inch, $7/16$-inch, $1/2$-inch, or 0.6 diameter seven-wire strand.
(T) _____ F _____ Reference _____

24. Batching of precast materials is by weight, although water and liquid admixtures can be batched by volume.
   (T)_____   F _____   Reference _____

25. Casting bed bulkheads are usually set with a space of 6 inches between them to facilitate subsequent operations.
   T _____   (F)_____   Reference _____

26. Precast prestressed units can be stored on the ground and stacked after curing, provided the surface is level.
   T _____   (F)_____   Reference _____

27. In long precast prestressing beds it is sometimes the practice to oil the tendons before they are placed in the forms.
   T _____   (F)_____   Reference _____

28. At detensioning, the tension in the prestressing tendons is transferred to the concrete, placing the concrete under compression.
   (T)_____   F _____   Reference _____

29. Long casting beds are not practical for producing many units of identical cross-section and strand pattern.
   T _____   (F)_____   Reference _____

30. It is essential that the shoring for posttensioned concrete be left in place until the stressing is completed.
   (T)_____   F _____   Reference _____

31. The greatest majority of forms for precast concrete are made of steel.
   (T)_____   F _____   Reference _____

32. The most widely used prestressing steel in building construction is the $1/2$" - 270K - stress-relieved-seven-wire strand.
   T _____   (F)_____   Reference _____

33. The minimum specified yield strength for Grade 270 prestressing steel is 270,000 psi.
   T _____   (F)_____   Reference _____

34. A small amount of rust on the surface of prestressing steel is beneficial to bond.
   (T)_____   F _____   Reference _____

35. Unbonded single-strand tendon systems are required to be protected against corrosion in accordance with ACI's "Specification for Unbonded Single-Strand Tendons (ACI 423.6)."
   T _____   F _____   Reference _____

36. For the installation drawing shown on page 270 of the *Concrete Manual*, strand stressing number (19) consists of one strand with an indicated elongation of $7\text{-}^1/_4$ inches.

   T _____ F _____ Reference _____

37. For the installation drawing shown on page 270 of the *Concrete Manual*, strand stressing number (34) is 36 feet in length.

   T _____ F _____ Reference _____

III Completion

38. When installing unbonded tendons, an inspector should check that the tendons are placed at the correct _____ and _____ elevations and that the profiles are _____ and correctly _____.
   Reference _____

39. In posttensioned concrete the tendons are placed _____ the reinforcing steel, electric conduit, and mechanical work.
   Reference _____

40. The placing drawings for precast concrete units are usually prepared by the _____ or the _____.
   Reference _____

41. Where space permits, on site precasting can be adopted for buildings where there are many _____ units.
   Reference _____

42. In any prestressing operation there is a small amount of slippage that develops as the _____ grip the _____ at the _____.
   Reference _____

43. Pretensioning is the method of prestressing in which the tendons are elongated _____ to the placement of _____, and posttensioning is the method of prestressing in which the tendons are elongated _____ the placement of _____.
   Reference _____

44. Prestressing reinforcing wire and strand is available in low-relaxation and _____. Low-relaxation prestressing wire has a lower steel-relaxation _____ and a higher _____ strength.
   Reference _____

45. Most precast concrete units have lifting hardware _____ in the concrete when the unit is _____. This hardware usually consist of an _____ and an _____ element.
   Reference _____

46. Posttensioned tendons have a grease applied to the strand which acts as a
_____ coating and a _____ between the strand and the
_____.
Reference _____

47. The modulus of elasticity of prestressing steel averages about _____ psi.
This can vary as much as _____ percent between lots.
Reference _____

CHAPTER 21
LIGHTWEIGHT AND HEAVYWEIGHT CONCRETE

Objectives: To give an introduction to the batching, mixing, handling, placing and finishing of lightweight and heavyweight concrete.

Lesson Notes: Lightweight and heavyweight concrete have many similarities to normal-weight concrete; however, each of these two classes of concrete has special requirements that must be followed if their intended purpose is to be met. Compare the grading requirements of lightweight concrete given in Table 21.2 with those for normal lightweight concrete given in Table 8.5.

Key Points:

- What is the normal weight of concrete used in engineered structures?
- Why use lightweight or heavyweight concrete?
- How is lightweight concrete obtained?
- Name the two general types of lightweight concrete.
- How are the two distinguished?
- Do the two ever overlap?
- What is the primary reason to use lightweight structural concrete?
- Name some of the advantages of structural lightweight concrete.
- List the natural and manufactured materials that are used as aggregates in lightweight concrete.
- Give the properties of lightweight aggregates for structural concrete.
- Describe in detail the two processes for manufacturing lightweight structural aggregates.
- What is the maximum absorption rate variation in the rotary kiln process?
- In which process is code or coal used?
- Which standard covers lightweight aggregates?
- Can the principles of normal-weight concrete proportioning be applied to lightweight concrete?
- What might be the difficulties?
- Explain the basics for the ACI Committee Standard 211 regarding proportioning of lightweight materials.
- What are the problems with lightweight concrete that need to be controlled?
- Give a brief description of the process of vacuum treatment of lightweight aggregate.
- How might the variations in specific gravity of particles be affected by water?
- Which affects the quality of lightweight concrete, active or free moisture?
- How is volumetric batching accomplished?
- Describe the appearance of fresh lightweight concrete.
- What slump is best for lightweight concrete slabs and structural elements?
- How should lightweight concrete be mixed in a truck mixer?
- What does a change in the unit weight mean?
- How is air content determined for lightweight concrete?
- What are the concerns regarding vibration of lightweight concrete?
- How is finishing of lightweight concrete different from that for normal-weight concrete?
- What is the density of lightweight insulating concrete?
- For what is lightweight insulating concrete used?
- Which types of aggregates are used for lightweight insulating concrete?
- What is perlite?
- What are the recommended proportions when using perlite or vermiculite?
- Give the water requirements for perlite and vermiculite.
- Describe the ways to mix insulating concrete at the site or in transit.
- What actions may cause insulating concrete to become denser?

- What is the most common use of insulating concrete?
- Briefly describe the methods of placing insulation concrete.
- Define "cellular concrete."
- Describe the two methods for making mechanically foamed cellular concrete.
- Where is heavyweight concrete most frequently used?
- Name the principal aggregates used for heavyweight concrete, and give a brief description of each.
- List the requirements for heavyweight concrete with regard to vibration, placing, mixing batch size and form construction.
- What is the intrusion method of placing concrete?
- How is proportioning accomplished with heavyweight concrete?
- How is heavyweight concrete affected by temperature?

CHAPTER 21—QUIZZES

I Multiple Choice

1. Which of the following is not one of the principle aggregates used in heavyweight concrete?
    a. barite
    b. granite
    c. limonite
    d. magnetite
    e. iron
    Response _____ Reference _____

2. Screeding and bullfloating operations must be kept to a minimum in the finishing process because of to the tendency of the aggregate to _____.
    a. segregate
    b. float to the surface
    c. absorb additional water
    d. sink to the bottom
    e. none of the above
    Response _____ Reference _____

3. In the kiln process of manufacturing aggregate, the material reaches a temperature of _____ degrees F.
    a.   800 to 1000
    b. 1000 to 1200
    c. 1200 to 1600
    d. 1600 to 1800
    e. 1800 to 2000
    Response _____ Reference _____

4. Structural lightweight concrete usually has a compressive strength in excess of _____ psi at 28 days.
    a. 1800
    b. 2000
    c. 2500
    d. 3000
    e. 3250
    Response _____ Reference _____

5. Aggregate for lightweight insulating concrete includes _____.
    a. limonite
    b. barite
    c. magnetite
    d. perlite
    e. all of the above
    Response _____ Reference _____

II True/False

6. Manufactured aggregates for structural lightweight concrete do not include clay and slate.

    T _____   F _____   Reference _____

7. Except for absorption factors, the principles of normal-weight concrete proportioning do not apply to lightweight concrete.

    T _____   F _____   Reference _____

8. In heavyweight concrete, segregation concerns are the same as for normal-weight concrete because the specific gravity is about the same.

    T _____   F _____   Reference _____

9. Natural aggregates used in structural lightweight concrete are normally smooth and round in shape, except for coated manufactured aggregates.

    T _____   F _____   Reference _____

10. The appearance of fresh lightweight concrete is similar to that of normal-weight concrete.

    T _____   F _____   Reference _____

III Completion

11. In the sintering process of manufacturing aggregates for structural lightweight concrete, the raw material is _____ and _____, then mixed with a _____ amount of pulverized _____ or _____.

    Reference _____

12. Cellular concrete contains bubbles of _____ or _____ that are formed in the plastic mortar with the porous structure _____ after the material hardens.

    Reference _____

13. Lightweight concrete in walls and columns should be consolidated by _____ using _____. Special care must be used to prevent _____.

    Reference _____

14. In lightweight concrete, differences in the amount of _____ water result from slight variations in the _____ of the particles, time of exposure to _____ and different mixes.

    Reference _____

15. One method of making mechanically foamed cellular concrete is to mix the cement, aggregate, _____ and _____ together in a _____ or _____ mixer.

    Reference _____

CHAPTER 22
SPECIAL CONCRETING TECHNIQUES

Objectives: To obtain a general awareness of the special concreting techniques of tilt-up construction, slipforms, lift slabs, placing concrete under water, preplaced aggregate concrete, vacuum concrete and shotcrete. To give an introduction to polymer, fiber-reinforced, refractory, sulfur, cellular and self-consolidating concrete, and controlled low-strength "backfill" material. Also, to provide a review of the architectural applications of concrete.

Key Points:

- Define "tilt-up construction."
- What is used as the casting platform for tilt-up construction?
- For what is tilt-up best suited?
- What is placed on the platform prior to placing concrete?
- What is the best type of bond breaker for tilt-up construction?
- What will happen if there are imperfections in or on the casting platform floor?
- Of what are side forms usually made?
- Name some finishes that can be applied while concrete is still plastic.
- Give two methods for setting tilt-up panels.
- When can a tilt-up panel be raised?
- How are pickup points determined?
- How can panels that need to be broken loose from the casting floor be moved without injury to the concrete?
- How are temporary braces attached?
- When are columns placed in tilt-up construction?
- In what ways are columns tied to the panels?
- What is a slipform?
- What structures are well suited to slipform construction?
- Describe how a slipform works.
- In vertically moving slipforms, what is the purpose of having a slight draft?
- How close to being plumb should a vertical slipform be?
- How is true vertical movement provided for a slipform?
- How is a level condition maintained on a vertical slipform?
- Can projections beyond the face of a vertical slipform be provided?
- What two methods are used to set doors and other openings in a vertical slipform?
- How important is planning to vertical slipform construction?
- What is the recommended slump of concrete used in vertical slipforms?
- Give the important considerations for vertical slipforms in the following areas: consolidation, placing delays and time constraints, finish, curing, and rate of slip.
- For what are horizontal slipforms used?
- Describe the operation of a horizontal slipform.
- Briefly describe the lift slab technique.
- What is the usual jacking rate of a lift slab?
- What two items are of special importance to lift slabs?
- Can concrete be placed in running water?
- Is the volume of cement usually increased when placing concrete in water? If so, how much and why?
- What admixtures are advantageous when placing concrete in water?
- What is the recommended slump for concrete placed under water?
- What methods of placement are used for concrete placed underwater?
- What is a tremie?

- How should a tremie be supported?
- How is the best end-control achieved?
- Describe the method that uses a cone valve to control concrete placement with a tremie.
- What criteria are followed for placing concrete with a tremie?
- What are the advantages and disadvantages of using a tremie?
- How should a bucket be operated when used to place concrete underwater?
- What is considered to be the best method for placing concrete? Describe this method.
- How is concrete placement with a pump accomplished?
- Why is it important to keep the discharge end of the pump submerged in the fresh concrete?
- Briefly describe the preplaced aggregate method.
- Where is the preplaced aggregate method used most frequently?
- What should be the aggregate size if regular concrete sand is used? If plaster sand is used?
- Which admixtures are used in preplaced aggregate concrete?
- Define "vacuum concrete."
- How is the vacuum process accomplished?
- Approximately how much water is removed during this process?
- Name the benefits of the vacuum process.
- Define "shotcrete."
- By what other name is shotcrete known?
- On what properties and performance is shotcrete dependent?
- Is shotcrete more economical? Explain.
- List the uses of shotcrete.
- Describe the dry-mix and wet-mix methods of preparing shotcrete.
- Compare and contrast the merits of each.
- Can shotcrete be used to repair concrete?
- How is the shotcrete procedure carried out in a variety of conditions and situations?
- What should be the pressure at the gun tank if the hose is 150 feet in length?
- What is rebound?
- Can rebound be reused? Explain.
- How is shotcrete finished and cured?
- Briefly describe how shotcrete is tested.
- What is a base plate?
- Describe two methods for setting base plates.
- How are the anchor bolts for a base plate set?
- What is the correct way to set the anchor bolt template?
- How thick can column bearing plates be?
- Can a column base ever be set in fresh concrete? Explain your answer.
- What is dry pack and how is it installed?
- Why add powdered aluminum to grout?
- Where might prebagged dry concrete be used?
- Define "polymer concrete."
- What are the two types of polymer concrete?
- Describe the polymer-impregnated process.
- Compare and contrast the polymer-impregnated and the polymer-portland cement processes to each other and conventional concrete.
- What is fiber-reinforced concrete?
- Name the types of fiber used in fiber-reinforced concrete.
- What are the common uses for each of these types of fiber?
- Give the recommended content percentage for each type of fiber.
- Where is refractory concrete used?

- Can refractory concrete be used for structural components?
- How durable is refractory concrete?
- List the types of aggregates used in refractory concrete.
- When is concrete classified as architectural?
- Give the fundamental requirements for all architectural concrete.
- How much finishing is applied to architectural concrete?
- Describe the four categories of architectural concrete.
- Why make a sample panel prior to placing architectural concrete?
- How is pigmented concrete mixed and placed?
- What special precautions must be taken when using pigmented concrete?
- How long should concrete age before paint is applied?
- How is portland cement paint applied and cured?
- What failures can occur when portland cement paint is applied incorrectly?
- List the other types of paints that can be used on concrete.
- Review Chapter 17 on exposed aggregate finishes. Describe the sand-bedding and aggregate transfer methods for preparing exposed aggregate.
- How many varieties of textures can be made for concrete surfaces, and what are the limitations?
- What does the term "rubbing" mean?
- Why and when is rubbing used?
- Describe in detail the rubbing process.
- How is a stoned or sand finish obtained?
- Approximately how thick should the grout be?
- What is grout cleaning?
- How old should concrete be before attempting treatment?
- Of what does grout for cleaning consist?
- Describe the cleaning process.
- Identify the precautions necessary when using white portland cement.
- Give six suggestions that will improve the quality appearance of white concrete.
- What effects do various aggregates have on white concrete?
- How do admixtures and pigments respond to white concrete?
- What are the considerations to be given to forms for white concrete?
- How are materials for white concrete batched?
- How might mixing time affect white concrete?
- With what types of surfaces is white cement used?
- How is finishing and curing of white concrete done?
- On what does roughness depend when sandblasting concrete?
- Will sandblasting remove surface lines?
- What type of aggregate is used in sandblasting?
- When is sandblasting usually done?
- How is tooling of a concrete surface accomplished?
- What is a bushhammer?
- How and for what is a bushhammer used?
- What is acid etching?
- On what does the final roughened surface of an acid-etched element depend?
- How is etching done at a precast plant?
- At what concentration is the acid?
- How is the acid applied?
- What precautions must be taken when acid etching?
- How old should concrete be before grinding is done?
- How is sulphur concrete produced?

- Define autoclaved cellular concrete (ACC).
- What are the principal ingredients in ACC?
- How is ACC manufactured?
- What is the primary construction use for ACC?
- Define "self-consolidating concrete (SCC)."
- Describe the primary use of SCC.
- Describe the J-ring test method for SCC.
- What is controlled low-strength material (CLSM)?
- What is the primary use of CLSM?
- What is "one precaution" in the use of CLSM?

CHAPTER 22—QUIZZES

I Multiple Choice

1. A type of construction where wall panels are cast in a horizontal position at the site is called _____.
   a. slipform
   b. lift slab
   c. shotcrete
   d. tilt-up
   e. none of the above
   Response _____ Reference _____

2. Finishers on horizontal slipforms make repairs and contraction joints from _____.
   a. an outrigger
   b. grade
   c. openings in the center of the form
   d. an apron
   e. none of the above
   Response _____ Reference _____

3. When shotcreting, the nozzle should be held uniformly about _____ feet away from the surface.
   a. 6
   b. 5
   c. 4
   d. 3
   e. 2
   Response _____ Reference _____

4. The four surface classes for concrete are _____.
   a. 1, 2, 3 and 4
   b. A, B, C and D
   c. integral, smooth, rough and treated
   d. unfinished, smooth, rough and semirough
   e. none of the above
   Response _____ Reference _____

5. When appearance is important, the recommended amount of white cement is about _____ pounds per cubic yard.
   a. 500
   b. 560
   c. 620
   d. 640
   e. none of the above
   Response _____ Reference _____

6. When using pumping to place concrete underwater, the discharge end must be kept continuously _____.
   a. submerged in the fresh concrete
   b. charged with water
   c. ahead of the concrete
   d. at the bottom of the element
   e. none of the above
   Response _____ Reference _____

7. When sulphur concrete is placed, the temperature of the concrete must be between _____ degrees F.
   a. 150 and 200
   b. 175 and 225
   c. 275 and 300
   d. 350 and 425
   e. none of the above
   Response _____ Reference _____

8. To make an expansive grout, powdered aluminum can be added in the amount of _____ to a sack of cement.
   a. 1 pound
   b. $1/_2$ pound
   c. 1 cupful
   d. 1 quart
   e. 1 teaspoonful
   Response _____ Reference _____

9. Columns between tilt-up panels may be bonded to the panel concrete with _____ cast in the panel and extending into the column.
   a. tie bars
   b. jacks
   c. stirrups
   d. rigging
   e. none of the above
   Response _____ Reference _____

10. If plaster sand is used in preplaced aggregate concrete, the coarse aggregate can be as small as _____ inch.
    a. 1
    b. $7/_8$
    c. $3/_4$
    d. $1/_2$
    e. $3/_8$
    Response _____ Reference _____

11. Concrete should be at least _____ days old before grinding the surface.
    a. 7
    b. 14
    c. 21
    d. 28
    e. 35
    Response _____ Reference _____

12. Concrete mixes for vertical slipforms should have a slump of between _____ inches.
    a. 4 and 6
    b. 3 and 6
    c. 4 and 8
    d. 2 and 6
    e. 2 and 4
    Response _____ Reference _____

13. Accurately setting anchor bolts for a base plate can be done by means of _____.
    a. reinforcing dowels
    b. set screws
    c. a template
    d. embedded nuts
    e. hooks or stirrups
    Response _____ Reference _____

14. The time and method of rubbing is stated in the _____.
    a. job specifications
    b. building code
    c. placing drawings
    d. rubbing manual
    e. curing schedule
    Response _____ Reference _____

15. Concrete conveyed through a hose in a stream of air and shot onto a surface at high velocity is known as _____.
    a. gunite
    b. vacuum concrete
    c. refractory concrete
    d. shotcrete
    e. polymer concrete
    Response _____ Reference _____

16. Sandblasting may cut as deep as _____ inch(es).
    a. $1/4$
    b. $1/2$
    c. $3/4$
    d. 1
    e. $1^1/2$
    Response _____ Reference _____

17. After curing, polymer-impregnated concrete can achieve compressive strengths of between _____ psi.
    a. 3000 to 5000
    b. 3000 to 8000
    c. 5000 to 18,000
    d. 5000 to 25,000
    e. none of the above
    Response _____ Reference _____

18. If a pure white concrete is specified, white sand and coarse aggregate is made by crushing white _____.
    a. quarzite
    b. limestone or quartz
    c. granite or mica
    d. marble or feldspar
    e. none of the above
    Response _____ Reference _____

19. In tilt-up construction, a _____ must first be placed on the casting floor.
    a. pickup point
    b. sack coat
    c. polymer
    d. epoxy resin
    e. bond breaker
    Response _____ Reference _____

20. Portland cement paint should have a creamy, thick consistency and should be applied with _____.
    a. a spray gun
    b. scrub brushes
    c. horse hair brushes
    d. sponges
    e. a wood float
    Response _____ Reference _____

21. Autoclaved cellular concrete is a porous material with a compressive strength between_____.
   a. 150 and 300 psi
   b. 300 and 1000 psi
   c. 300 and 1500 psi
   d. 1000 and 3000 psi
   e. 1500 and 2500 psi
   Response _____ Reference _____

22. A self-compacting structural concrete that can be proportioned to flow into tight and inaccessible spaces is termed_____.
   a. autoclaved aerated concrete
   b. controlled low-strength concrete
   c. polymer concrete
   d. self-consolidating concrete
   e. shotcrete
   Response _____ Reference _____

23. A well-proportioned self-compacting concrete will have a slump diameter of approximately_____.
   a. 3 inches
   b. 9 inches
   c. 15 inches
   d. 30 inches
   e. none of the above
   Response _____ Reference _____

24. Controlled low-strength material is _____.
   a. a flowable fill material
   b. a porous building material
   c. self-leveling concrete
   d. very flowable concrete
   e. all of the above
   Response _____ Reference _____

25. The "J-ring" is a modified slump test used to measure unblocked flow of_____.
   a. autoclaved aerated concrete
   b. controlled low-strength concrete
   c. fiber-reinforced concrete
   d. self-supporting concrete
   e. none of the above
   Response _____ Reference _____

26. A freshly mixed batch of pervious concrete has a_____.
    a. high cement-past content
    b. low void content
    c. high fine aggregate content
    d. very low slump
    e. all of the above
    Response _____ Reference _____

27. Ultra-high performance concrete provides compressive strengths up to about_____ psi.
    a. 5000
    b. 10,000
    c. 15,000
    d. 20,000
    e. 29,000
    Response _____ Reference _____

28. Ultra-high performance concrete provides a material that is very_____.
    a. durable
    b. ductile
    c. high strength
    d. impermeable
    e. all of the above
    Response _____ Reference _____

II True/False

29. Raising of a lift slab is accomplished by means of jacks mounted on top of the building columns.
    T _____   F _____   Reference _____

30. There are four basic shotcreting processes: dry-mix, wet-mix, pneumatic and injected.
    T _____   F _____   Reference _____

31. Grout for cleaning consists of one part cement with one and one-half to two parts fine sand that passes a 16 mesh screen.
    T _____   F _____   Reference _____

32. Steel-fiber-reinforced concrete that has about a five percent fiber content by volume is considered the upper limit.
    T _____   F _____   Reference _____

33. Samples of shotcrete for strength tests are made by filling a 6 inch by 12 inch cylinder directly from the nozzle.
    T _____   F _____   Reference _____

34. After proper curing, refractory concrete can be heated up immediately at a rapid rate.

    T _____   F _____   Reference _____

35. A bushhammer consists of a flat-faced tool that fits into a chipping gun.

    T _____   F _____   Reference _____

36. The vacuum concrete process applies a vacuum to a fresh concrete surface through a medium of special permeable form liners or pads.

    T _____   F _____   Reference _____

37. The sand-bedding technique, used in exposed aggregate concrete, allows a depth of exposure of up to 4 inches.

    T _____   F _____   Reference _____

38. Among the causes of color variation in white concrete are different brands of cement, different forming materials, different slumps and variations in curing.

    T _____   F _____   Reference _____

39. Vertical slipforms consist of an inside and outside form made of sheet steel. The outside form extends above the inside form about 6 inches.

    T _____   F _____   Reference _____

40. Sack rubbing is done to fill in or cover rock pockets or honeycombing defects.

    T _____   F _____   Reference _____

41. Slipforms must have a steady supply of concrete available and placement made so that there is not more than an hour's delay between lifts.

    T _____   F _____   Reference _____

42. The principal advantage of using a tremie to place concrete underwater is that dewatering of the foundation area is unnecessary.

    T _____   F _____   Reference _____

43. The size and location for pickup points on a tilt-up panel are determined by its size, weight, compressive strength and unit weight.

    T _____   F _____   Reference _____

44. Autoclaved cellular concrete is a nonstructural lightweight precast concrete building material.

    T _____   F _____   Reference _____

45. Autoclaved cellular concrete (ACC) can be used for structural applications if properly reinforced.

    T _____   F _____   Reference _____

46. Self-consolidating concrete is proportioned with about the same amount of mixing water as conventional concrete.
   T _____   F _____   Reference _____

47. Autoclaved cellular concrete is a special type of lightweight precast prestressed concrete building material.
   T _____   F _____   Reference _____

48. Controlled low-strength material requires some vibration for adequate consolidation.
   T _____   F _____   Reference _____

49. Individual ACC building elements are joined together by embedded dowels or ties.
   T _____   F _____   Reference _____

50. Self-consolidating concrete is proportioned to flow between and around reinforcement without requiring any vibration.
   T _____   F _____   Reference _____

51. Pervious concrete is very high impermeable concrete that drains quickly.
   T _____   F _____   Reference _____

52. Pervious concrete resembles "popcorn."
   T _____   F _____   Reference _____

53. The void structure of pervious concrete allows water to pass through and percolate into the ground.
   T _____   F _____   Reference _____

54. The addition of plastic fibers in a concrete mixture will require more water to maintain a specified slump.
   T _____   F _____   Reference _____

III Completion

55. Part of the wet-mix shotcrete process is that all ingredients, including _____, are thoroughly mixed together, placed in the delivery equipment _____ and conveyed by _____ to a nozzle.
   Reference _____

56. Glass-fiber-reinforced concrete is manufactured by a spray-up process that feeds a continuous strand of glass fiber into a compressed-air-powered _____, where it is cut into _____ and combined with a _____ and _____ slurry.
   Reference _____

57.   Etching can be done as soon as _____ days(s) after placing concrete, and all comparable areas should be etched at about the same _____.
        Reference _____

58.   Methods for placing concrete underwater include the use of _____, _____ and _____.
        Reference _____

59.   After a base plate has been adjusted to the correct position, the space underneath is filled with _____ or _____.
        Reference _____

60.   When using pigments to color concrete, only pure metallic _____ should be used, in an amount determined by _____.
        Reference _____

61.   When repairing old concrete with shotcrete, all old unsound material must be _____, corroded steel must be _____, and reinforcing securely _____ or _____ in place.
        Reference _____

62.   When acid is applied to a concrete surface, the acid reacts with the _____ and will also attack _____ and _____ aggregate.
        Reference _____

63.   Compared to untreated concrete, polymer-impregnated concrete has strength values _____ times greater, improved resistance to _____ and _____, increased resistance to _____ attack, improved _____ resistance and _____ water absorption.
        Reference _____

64.   When using a bucket to place concrete underwater, the bucket should be lowered _____ while underwater and should not be opened until the bucket contacts _____ concrete.
        Reference _____

CHAPTER 23
WATERPROOFING AND DAMPPROOFING

Objectives: To give an introduction to the dampproofing and waterproofing of concrete and to some of the available materials and methods used to achieve this.

Lesson Notes: There are many materials and methods available for dampproofing and waterproofing of concrete. Care must be taken to follow all the manufacturer's directions explicitly to obtain an acceptable and lasting seal. There are also many new products not mentioned in the text that are effective in the repair of leaks in existing structures.

Key Points:

- Describe the two ways that water passes through concrete.
- How is each of these ways controlled?
- What can contribute to the problem of maintaining a water-tight structure?
- Review permeability in Chapter 5 and waterproofing in Chapter 9.
- List the most important points given in these two chapters.
- Of what materials do surface treatments consist?
- Give one effective method of providing protection of porous concrete under low water pressure.
- Describe some ways to provide drainage away from concrete walls.
- Is waterproofing always required?
- What are the three primary requirements for waterproofing or dampproofing concrete?
- List the types of materials used to waterproof concrete.
- Where is waterproofing required?
- When is special care needed in all of the systems listed in this chapter?
- To what can most failures be traced?
- List the concerns associated with the installation of the membrane.
- Of what does an elastomeric membrane consist?
- How is it applied and what care must be taken during installation?
- How does preformed sheet elastomeric membrane differ?
- How are single-component liquids applied?
- What is the minimum number of plies when using a bituminous membrane for waterproofing?
- Describe the conditions for application of this type of system and how it is installed.
- When using plaster to waterproof concrete, how is it applied?
- How is sheet lead used to waterproof concrete?
- Compare and contrast each of the preceding methods and list their advantages and disadvantages.
- When is dampproofing appropriate?
- What is the difference between dampproofing and waterproofing?
- Can treatments for dampproofing be substituted for waterproofing? Is the reverse also true?
- Give a detailed description of how to seal a leaking structure subject to a hydrostatic head.
- Is Type III cement a good material for this purpose?

CHAPTER 23—QUIZZES

I Multiple Choice

1. When a waterproofing system fails, the problem can usually be traced to _____.
    a. improper construction
    b. material breakdown
    c. faulty materials
    d. temperature fluctuations
    e. all of the above
    Response _____   Reference _____

2. Quick-setting cement can be made by mixing Type III cement with _____.
    a. perlite
    b. aluminous cement
    c. calcium chloride
    d. magnesium sulfate
    e. none of the above
    Response _____   Reference _____

3. Walls in basements should have surface water drain by sloping the ground away from the structure about $1/2$ inch in _____ feet.
    a. 5
    b. 10
    c. 15
    d. 20
    e. 25
    Response _____   Reference _____

4. To ensure the watertightness of concrete it should be wet cured for at least _____ days.
    a. 3
    b. 6
    c. 7
    d. 14
    e. 28
    Response _____   Reference _____

5. Modified polyurethanes that are applied directly to the concrete from a can and spread with a notched squeegee are known as _____.
    a. sheet membrane
    b. bituminous membrane
    c. elastomeric membrane
    d. single-component liquid
    e. none of the above
    Response _____   Reference _____

II True/False

6. Waterproofing materials cannot be used to dampproof a structure.

   T _____   F _____   Reference _____

7. Plaster used to waterproof a structure is applied either by hand or machine in three coats, each about $^3/_8$ inch thick.

   T _____   F _____   Reference _____

8. A waterproof membrane should be protected as soon as it has been installed, and if the membrane is punctured it can be replaced by applying a patch of the material.

   T _____   F _____   Reference _____

9. Outdoor pools are sealed with a membrane of sheet lead that is placed prior to placing concrete.

   T _____   F _____   Reference _____

10. There are usually two plies of bituminous membrane applied to an exterior vertical surface.

    T _____   F _____   Reference _____

III Completion

11. Manufacturers of bituminous membrane usually specify that prior to application the concrete is _____, _____ and _____. Also, all surface voids must be _____ with _____ and all fins and irregularities _____.

    Reference _____

12. Leaks can be repaired by removing _____ concrete and _____, and cracks should be _____. A good proprietary material is then applied, starting from the _____ and working to the _____ point.

    Reference _____

13. Bituminous coatings consist of _____ or _____ layers of bitumen, mopped on either _____ or _____. Cold-applied bituminous coatings can be reinforced with _____, _____ or other inert fibers.

    Reference _____

14. Waterproofing is required below _____ where groundwater is present against _____ and _____, and above grade wherever protection is required against the _____.

    Reference _____

15. To assure watertight impermeable concrete, aggregates should be _____ and of _____, and sand particles should be _____.

    Reference _____

CHAPTER 24
INTRODUCTION TO INSPECTION

Objectives: To give an overview of the responsibilities and authority of building, special and quality control inspectors.

Lesson Notes: The job of the inspector is probably the most difficult of all of the members of the construction team. He or she must understand and apply all of the various tests, procedures, code requirements and specifications related to each individual project. He or she must know not only the exact wording of each of these but the intent as well, insofar as each project presents its own unique problems and conditions.

Key Points:

- Why is the team concept important in concrete?
- List each of the team players and their roles in providing a good quality product.
- Define "inspection."
- Who might the inspector represent?
- What is the objective of inspection?
- How is this objective met?
- What is the typical arrangement for inspection when a building permit is required?
- Why is it not recommended to award a contract for inspection services to the lowest bidder?
- How is inspection of government projects done?
- What can be the advantages to contractors who provide their own inspection staff?
- Who should employ the testing or inspection staff for the owner or building official?
- List the qualities of a good inspector.
- To whom should the inspector give suggestions and instructions?
- How should an inspector act when there is a difference of opinion between an inspector and a supervisor?
- How should the supervisor support the inspector?
- When a permit is required, who is the primary inspector?
- Describe the responsibilities of a special inspector.
- Upon what is the general building code based?
- Is the building code the only document with which the inspector must be familiar? Explain.
- List the primary sources that should guide the inspector?
- What are considered to be the secondary sources that guide the inspector?
- What is the first duty of an inspector when assigned a project?
- What documents are needed to perform this duty?
- List the duties of the inspector.
- When may a field laboratory be required?
- What equipment does a testing agency usually provide on the job site?
- What is the responsibility of the inspector in relation to the materials used?
- On what is a preliminary approval based?
- Who gives this approval?
- Which materials are usually tested at the manufacturer?
- What should accompany approved materials?
- When can rejected materials be used on a site?
- How should rejected materials be disposed of?
- When are retests of rejected materials appropriate?
- How should the inspector be involved in job safety?
- What questions should be asked when inspecting materials?

- Describe how cement is delivered, identified and inspected.
- When may an inspector be involved in approval of aggregates?
- What should an inspector assigned to a batch plant inspect?
- What types of tests might an inspector perform on aggregates?
- Describe some methods for testing the moisture content of aggregates.
- What does the inspector check when inspecting reinforcing steel both on and off the job site?
- When can alternate materials be used on a job site?

CHAPTER 24—QUIZZES

I Multiple Choice

1. When special inspection is required, the special inspector shall be in the employ of the _____.
   a. contractor
   b. subcontractor
   c. owner
   d. building official
   e. none of the above
   Response _____ Reference _____

2. The primary code resource document for concrete construction is the _____.
   a. general building code
   b. ACI building code
   c. ACI manual of concrete construction
   d. ASTM standard specifications
   Response _____ Reference _____

3. When a permit is required, the inspector employed by the building official is the _____ inspector.
   a. primary
   b. secondary
   c. special
   d. additional
   e. none of the above
   Response _____ Reference _____

4. Safety and accident prevention on the job site are the responsibility of the _____.
   a. owner
   b. inspector
   c. architect
   d. engineer
   e. contractor
   Response _____ Reference _____

5. There are _____ primary sources of authority that guide the inspector.
   a. one
   b. two
   c. three
   d. four
   e. five
   Response _____ Reference _____

6. Job specifications usually permit the use of alternative materials of equal quality, provided necessary test reports and other pertinent information is submitted for approval by the _____.
   a. building official
   b. engineer
   c. architect
   d. owner
   e. engineer and owner
   Response _____ Reference _____

7. At the time of use, cement should contain no lumps that cannot be broken by _____.
   a. a hammer
   b. crushing
   c. light pressure between the fingers
   d. the aggregate
   e. none of the above
   Response _____ Reference _____

8. When required, test samples of reinforcing steel should be chosen at random from the lot. Samples should be at least _____ inches long.
   a. 12
   b. 18
   c. 20
   d. 24
   e. 30
   Response _____ Reference _____

9. Which of the following is not provided by the testing and/or inspection agency?
   a. slump cone
   b. on-site storage
   c. scoop or shovel
   d. cylinder molds
   e. air content meter
   Response _____ Reference _____

10. The new single model code organization in the United States is called_____.
    a. Building Officials and Code Administrators International (BOCA)
    b. International Code Council (ICC)
    c. International Conference of Building Officials (ICBO)
    e. Southern Building Code Congress International (SBCCI)
    d. none of the above
    Response _____ Reference _____

11. The_____edition of the *International Building Code®* is the current edition of the IBC®.
a. 2000
b. 2001
c. 2002
d. 2003
e. 2004
Response _____ Reference _____

12. For quality and testing of materials used in concrete construction, the inspector needs to refer to_____.
a. *International Building Code*
b. ASTM standard specifications
c. ACI Standard 318 for structural concrete
d. ACI Standard 301 specifications for structural concrete
e. All of the above
Response _____ Reference _____

II True/False

13. Inspection is the review of a contractor's work to make sure that specifications, drawings and codes are being followed.
T _____ F _____ Reference _____

14. Cement is rarely furnished to a job site, because practically all concrete comes from a commercial ready-mix manufacturer.
T _____ F _____ Reference _____

15. A special inspector is required to be on site only while concrete is being placed.
T _____ F _____ Reference _____

16. One of the first duties of an inspector is to become familiar with the job requirements that pertain to inspection.
T _____ F _____ Reference _____

17. Admixtures, curing compounds, joint fillers and similar materials are usually accepted on the manufacturer's certification.
T _____ F _____ Reference _____

18. Rejected materials should be disposed of, modified or regenerated.
T _____ F _____ Reference _____

19. Each load of approved materials should be accompanied by a tag or card of identification issued by the testing laboratory.
T _____ F _____ Reference _____

20. When the specifications require a particular material, substitution of a different material, even if of equal quality, is never allowed.
    T _____   F _____   Reference _____

21. The approval of materials is usually the responsibility of the on-site inspector.
    T _____   F _____   Reference _____

22. The "Special Inspection" requirements for building construction are addressed in Chapter 17 of the *International Building Code*.
    T _____   F _____   Reference _____

23. Four- by eight-inch cylinder molds are never permitted for final evaluation and acceptance of structural concrete.
    T _____   F _____   Reference _____

24. The *International Building Code* adopts by reference ACI Standard 301 for structural concrete to regulate concrete design and construction.
    T _____   F _____   Reference _____

III Completion

25. Of the several methods for obtaining the moisture content of an aggregate, the most common method is to _____ the aggregate in and _____ or over a
    _____.
    Reference _____

26. Although cement is manufactured under close _____ and rarely fails to meet _____, wide fluctuations in the cement's _____ may still exist.
    Reference _____

27. A qualified inspector should be thoroughly familiar with the applicable _____ Standards and Chapters _____ through _____ of the ACI 318 Standard.
    Reference _____

28. To obtain approval of a material, supporting data should be supplied that contain the history and _____ record as well as typical _____, _____ or shop _____, including those by an _____ testing laboratory.
    Reference _____

29. When an inspector is assigned a project, one of his or her first tasks is to become familiar with the _____ requirements and the _____.
    Reference _____

30. The inspector should give _____ and _____ relative to the acceptance or rejection of construction or materials to the contractor or producer, not the _____.
Reference _____

31. An inspector at a batch plant should check the aggregate and have all _____, _____ or other _____ removed.
Reference _____

CHAPTER 25
INSPECTION OF CONCRETE CONSTRUCTION

Objectives: To build on the information provided in Chapter 24 by deepening the understanding of the duties and responsibilities of the inspector, from preliminary arrangements to the final product.

Lesson Notes: One of the most important aspects of an inspector's job is to keep accurate records and reports. When good records and reports are kept, problems and questions that arise afterward can be addressed with facts instead of speculation.

Key Points:

- List the factors that determine the amount and extent of inspection.
- List the elements of foundations, floors and roofs, walls and columns, and slabs-on-ground that may be subject to inspection.
- What may be included in the preliminary inspection of the following: general approval of materials, cement, aggregates, special aggregates, admixtures, concrete mixture, ready-mix plant and laboratory?
- How can deficiencies in concrete be minimized at the plant?
- After the preliminary inspections, what are the three stages of inspection?
- What does each of these stages cover?
- What is the most frequent method of batching and mixing?
- What items need to be inspected at the time of proportioning and mixing?
- List the duties of the plant inspector at the beginning of each day.
- Is the inspector's only job during concrete placement simply to observe? Explain.
- At what point does an inspector take samples and perform tests?
- In the first stage of inspection, briefly describe the items to be inspected that relate to: foundations and excavations, forms, reinforcement, embedded items and site facilities.
- What items are to be inspected at the batch plant during the concreting phase?
- List the types of inspection that should occur during mixing, delivery, handling and placing of concrete.
- What should be inspected during jointing and finishing?
- Curing, removal of forms and repair of defects are part of the final inspection stage. Give a list of items to be inspected in each of these areas.
- After studying Section 25.6, make a list of inspection concerns for each of the following: prestressed concrete, hot and cold weather concreting, lightweight and heavyweight concrete, tilt-up construction, slipforms, placing concrete under water, shotcrete, architectural concrete, vacuum concrete, and fastening of base plates and preplaced aggregate concrete. If necessary, review the previous chapters.
- Why is the keeping of accurate records and reports necessary?
- When is a narrative report done?
- When an inspector is assigned numerous jobs, what items should his or her diary include?
- Define "special inspection."
- Why are special inspectors needed?
- Describe the role of the special inspector with relation to the enforcement agency.
- List the types of work related to concrete that are required to have special inspection.
- Give the general areas of responsibility and the qualifications of the special inspector.
- Review the job task analysis given in Table 25.1. Describe how each of the tasks might impact a project.
- Who approves a fabrication plant?
- What are the four general qualifications for fabricators?

173

CHAPTER 25—QUIZZES

I Multiple Choice

1. Of the following, which is not part of the first stage of inspection?
    a. steel grade and size
    b. soil compaction
    c. strength tests
    d. form stability
    e. adequate lighting
    Response _____ Reference _____

2. The general building code typically requires that the special inspector be employed by the _____.
    a. building official
    b. owner
    c. contractor
    d. subcontractor
    e. none of the above
    Response _____ Reference _____

3. The most important part in the approval of a fabrication plant is _____ by an approved quality control agency.
    a. independent inspection
    b. testing
    c. supervision
    d. sampling
    e. sampling and testing
    Response _____ Reference _____

4. A concrete inspection log should contain _____.
    a. strength specimen results
    b. unusual placing delays
    c. the number of workers
    d. ready-mix drum rotations
    e. all of the above
    Response _____ Reference _____

5. Ready-mix trucks should be checked by the inspector to verify that _____.
    a. engines are operational
    b. mixing water pump is adequate
    c. drums and chutes are clean of cement
    d. mixing blades are worn
    e. all of the above
    Response _____ Reference _____

6. A special inspector supplements inspections provided by the building official with
   _____ inspections to help ensure that construction complies with the code.
   a. partial
   b. periodic
   c. overtime
   d. continuous
   e. any of the above
   Response _____ Reference _____

7. Which of the following is not part of the preliminary arrangements?
   a. approving aggregates
   b. checking forms for line and grade
   c. calibrating scales and batchers
   d. preparing mix designs
   e. rejecting unsuitable materials
   Response _____ Reference _____

8. Which of the following is not part of inspection during the final stage of concreting?
   a. applying curing compound
   b. repairing rock pockets
   c. timely removal of forms
   d. installing construction joints
   e. filling tie rod holes
   Response _____ Reference _____

9. According to the *International Building Code*, continuous special inspection is
   required for_____.
   a. bolts installed in concrete
   b. concrete placement
   c. prestressing tendon stressing
   d. sampling fresh concrete
   e. all of the above
   Response _____ Reference _____

10. According to the *International Building Code*, periodic special inspection is permitted
    for_____.
    a. anchors embedded in concrete
    b. grouting prestressing tendons
    c. placing prestressing tendons
    d. welding reinforcing steel
    e. none of the above
    Response _____ Reference _____

II True/False

11. The second stage of inspection includes verifying the size, location and grade of the reinforcing steel.
     T _____   F _____   Reference _____

12. Areas of inspection of prestressed concrete include chamfer trip placement, correct steel size, detensioning and hardware placement.
     T _____   F _____   Reference _____

13. In building construction, the inspector is not called upon to review nonstructural elements of a building, except at the request of the building official.
     T _____   F _____   Reference _____

14. A special inspector should notify the building official and designer when discrepancies are not corrected.
     T _____   F _____   Reference _____

15. It is important for the inspector to maintain accurate and complete reports, but it is not necessary to include weather conditions and visitors to the job site.
     T _____   F _____   Reference _____

16. The general building code typically states that the fabricator's facility and personnel must be verified by an approved inspection or quality control agency.
     T _____   F _____   Reference _____

17. Repairs of rock pockets should be made as early as possible because it is easier to work on green concrete.
     T _____   F _____   Reference _____

18. It is normal practice to sample concrete and perform tests at the point of placement after all water has been added and while concrete is being discharged.
     T _____   F _____   Reference _____

19. Special considerations for tilt-up construction include applying parting compound, watching for rebound and avoiding sudden jerks when lifting.
     T _____   F _____   Reference _____

20. According to the *International Building Code*, periodic special inspection is permitted for in-situ concrete testing.
     T _____   F _____   Reference _____

21. According to the *International Building Code*, periodic special inspection is permitted for testing of fresh concrete.
     T _____   F _____   Reference _____

22. Types of concrete work requiring special inspection are detailed in Table 1704.4 of the *International Building Code*.
    T _____   F _____   Reference _____

23. The International Code Council offers a certification for reinforced concrete and prestressed concrete special inspector.
    T _____   F _____   Reference _____

24. Special inspection is always required for precast prestressed concrete building elements.
    T _____   F _____   Reference _____

25. Special inspection is always required for posttensioned prestressed concrete building elements.
    T _____   F _____   Reference _____

III Completion

26. The second stage of inspection of concrete occurs during the actual _____, _____ and _____ of the concrete, and extends through the _____ period.
    Reference _____

27. Although inspection may not be required, a _____ is usually necessary for concrete jobs, regardless of _____.
    Reference _____

28. The special inspector is responsible for furnishing _____ to the building official and observing the work for compliance with approved _____ and _____.
    Reference _____

29. In addition to verifying that applicants are technically competent, the building official should verify that applicants have related work experience and are aware of local code _____, _____ and _____.
    Reference _____

30. When inspecting prestressed concrete, an inspector should check the _____ of the steel stressing ram and the stressing _____.
    Reference _____

31. When checking reinforcing, an inspector should verify bar laps for _____ and bar bends for minimum diameter, _____ and _____.
    Reference _____

32.   Inspection of heavy-duty floors should include screeding, _____ and
       _____, troweling, wearing curse and special _____.
       Reference _____

33.   During concrete placement, the inspector should confirm that the _____
       indicates the correct mixture, the concrete is _____ and the mix is used
       within the specified _____.
       Reference _____

CHAPTER 26
QUALITY CONTROL

Objectives: To define quality control and its application to concrete construction.

Key Points:

- When can the principles of quality control be applied?
- Define "quality control."
- What are some of the primary areas in which quality control can be applied to construction?
- How can a concrete product manufacturer ensure that its products will perform acceptably for their intended purpose?
- How are the materials used in construction evaluated?
- What must occur to make the best use of available materials?
- Who is responsible for quality control of concrete?
- What benefit does an owner obtain from quality control?
- What is needed for quality control to succeed?
- How is this accomplished?
- What is the difference between quality control and acceptance sampling?
- How have recent advances in technology aided statistical quality control (SQC)?
- What information is provided by statistical quality control?
- On what is statistical quality control based? Explain briefly.
- What is a standard deviation?
- In what two ways is a standard deviation expressed?
- Define "coefficient of variation."
- What is the relationship between psi and compressive strength?
- What leads to the greatest uniformity in the quality of concrete?
- In the area of concrete quality control, why are rigid numerical limits unrealistic for contractors and inspectors?
- What is the best index of concrete quality?
- What test is used to determine concrete strength?
- What accounts for the differences in strength of test cylinders?
- Do low strength results in some cylinders mean that construction quality is jeopardized?
- Of what does a strength test consist?
- Is there an absolute minimum strength for concrete? Why or why not?
- What is a good index of the quality of concrete?
- Is the inspector expected to be able to make quality control computations?
- How would you arrive at an average in SQC? (Refer to Table 26.1.)
- How can 28-day results be determined based on seven-day strength curves?
- What action might be required to correct deficiencies in concrete quality?
- In the field control of concrete, what coefficient indicates good control? Fair control? Poor control?
- What is the standard procedure for computing the coefficient of variation and the standard deviation when a computer is not available?
- Why is it important for an inspector to understand the significance of statistical quality control?
- Does quality control result in added cost for the contractor?
- What section of the ACI 318 Standard states the requirements for concrete quality? On what are these requirements based?

CHAPTER 26—QUIZZES

I Multiple Choice

1. If quality control is to succeed, there must be a rational system for analyzing the results of _____.
    a. research
    b. tests
    c. samples
    d. SQC
    e. all of the above
    Response _____ Reference _____

2. An evaluation is possible to determine probable 28-day strengths from seven-day strength tests by using _____.
    a. strength averaging
    b. known mix designs
    c. statistical analysis
    d. a control chart
    e. all of the above
    Response _____ Reference _____

3. In general, strength is a good index of concrete _____.
    a. quality
    b. durability
    c. workability
    d. tensile strain
    e. uniformity
    Response _____ Reference _____

4. _____ is a measure of variation derived mathematically from test results.
    a. Standard deviation
    b. Range
    c. Average
    d. Coefficient of variation
    e. none of the above
    Response _____ Reference _____

5. The total number of test values under consideration is called the _____.
    a. range
    b. mean
    c. population
    d. deviation
    e. numeric average
    Response _____ Reference _____

6.   The calculated standard deviation (s = 353 psi) illustrated in Tables 26.3 and 26.4,
        for the column concrete with a specified strength of 4000 psi,
        represents_____.
        a. excellent quality control
        b. good quality control
        c. fair quality control
        d. poor quality control
        e. unacceptable quality control
        Response _____ Reference _____

7.   If a local ready-mix producer is proposing to use strength data with a standard
        deviation of 390 psi to bid on a project that requires concrete with a specified
        strength of 3500 psi, the required average strength used as the basis for selecting
        concrete mix proportions for the specified 3500 psi concrete should be_____.
        a. 3500 psi
        b. 3900 psi
        c. 4000 psi
        d. 4100 psi
        e. 4700 psi
        Response _____ Reference _____

II True/False

8.   Quality control is a system by which construction is controlled by scientific methods
        rather than chance.
        T _____   F _____   Reference _____

9.   The inspector is not usually called upon to make computations on the job site;
        however, he or she should know and understand the significance of the statistical
        values used, and thus how well the job is being controlled.
        T _____   F _____   Reference _____

10.   A slump test does not lend itself to the precision of measurement that a strength test
        does, and the results of the analysis ordinarily are not as meaningful.
        T _____   F _____   Reference _____

11.   Quality control is a relatively new concept with regard to products manufactured at a
        permanently located factory or mill.
        T _____   F _____   Reference _____

12.   To obtain accurate information, the results of a small number of tests should be
        presumed to be representative of the concrete produced.
        T _____   F _____   Reference _____

III Completion

13.   Computer programs allow a continuing analysis that provides up-to-the-minute information on _____, aggregate sieve _____, _____equivalents and any other test done on a _____ basis.
        Reference _____

14.   Statistical methods provide the best basis for analyzing test results, determining potential _____ and _____, and expressing _____ in the most useful form.
        Reference _____

15.   When writing specifications, it is more realistic to base probabilities on statistical methods and permitting a certain _____ of strength tests _____ than specified _____ strength.
        Reference _____

16.   Quality control _____ cost money, and the potential _____ are substantial.
        Reference _____

17.   The primary function of compression tests is to serve as a measure of the _____ and _____ of concrete. The magnitude of variations in strength of concrete test specimens depends on how well the _____, concrete _____and tests are _____.
        Reference _____

ANSWER KEYS

Chapter 1—Fundamentals of Concrete

1.	Sec.	1.1	b
2.	Sec.	1.8	c
3.	Sec.	1.3	b
4.	Sec.	1.2	a
5.	Sec.	1.1	e
6.	Sec.	1.5	T
7.	Sec.	1.8	F
8.	Sec.	1.7	F
9.	Sec.	1.1	T
10.	Sec.	1.2	T
11.	Sec.	1.3	green
12.	Sec.	1.6	durability
13.	Sec.	1.7	expansion, contraction, destructive solutions
14.	Sec.	1.1	gypsum
15.	Sec.	1.1	rotary kiln

Chapter 2—The Fresh Concrete

1.	Sec.	2.8	a
2.	Sec.	2.2	a
3.	Sec.	2.5	d
4.	Sec.	2.4	d
5.	Sec.	2.2	c
6.	Sec.	2.1	T
7.	Sec.	2.1	F
8.	Sec.	2.4	F
9.	Sec.	2.5	F
10.	Sec.	2.6	T
11.	Sec.	2.8	unit weight, bleeding
12.	Sec.	2.7	unit weight
13.	Sec.	2.1	consolidation, compaction
14.	Sec.	2.2	pavements, mass concrete, precast
15.	Sec.	2.6	bleeding

Chapter 3—The Strength of Concrete

1.	Sec.	3.3	d
2.	Sec.	3.11	b
3.	Sec.	3.13	e
4.	Sec.	3.11	b
5.	Sec.	3.15	d
6.	Sec.	3.11	b

7.	Sec.	3.7	b
8.	Sec.	3.11	a
9.	Table	3.1	d
10.	Sec.	3.2	a
11.	Fig.	3-2	d
12.	Sec.	3.4	c
13.	Sec.	3.5	c
14.	Sec.	3.7	b
15.	Sec.	3.11	b
16.	Fig.	3-7	c
17.	Sec.	3.13	d
18.	Sec.	3.14	b
19.	Table	3.5	c
20.	Sec.	3.2	c
21.	Sec.	3.5	T
22.	Sec.	3.9	F
23.	Sec.	3.13	T
24.	Sec.	3.17	T
25.	Sec.	3.2	T
26.	Sec.	3.15	T
27.	Sec.	3.17	F
28.	Sec.	3.11	T
29.	Sec.	3.15	F
30.	Sec.	3.14	slowed
31.	Sec.	3.4	modulus of rupture, third, 6 by 6
32.	Sec.	3.13	high-early, admixtures, heat of hydration, high-temperature, rapid setting cements
33.	Sec.	3.11	2.25, one and one-half
34.	Sec.	3.9	swiss hammer, windsor probe

Chapter 4—The Durability of Concrete

1.	Sec.	4.10	e
2.	Sec.	4.3	c
3.	Sec.	4.1	d
4.	Sec.	4.5	b
5.	Sec.	4.2	b
6.	Sec.	4.4	c
7.	Sec.	4.3	d
8.	Sec.	4.11	F
9.	Sec.	4.3	T
10.	Sec.	4.11	T
11.	Sec.	4.1	T
12.	Sec.	4.3	F
13.	Sec.	4.12	F
14.	Sec.	4.2	T
15.	Sec.	4.8	Chamfers, fillets

16.	Sec.	4.6	nonbreaking, breaking, broken
17.	Sec.	4.9	hydraulic, lowering
18.	Sec.	4.3	resistant, barrier
19.	Sec.	4.1	materials, physical propertied, exposure conditions, loads imposed, design, practices
20.	Sec.	4.3	Ammonium, ammonia, hydrogen, acid

Chapter 5—Volume Changes and Other Properties

1.	Sec.	5.11	c
2.	Sec.	5.1	c
3.	Sec.	5.1	d
4.	Sec.	5.11	a
5.	Sec.	5.8	c
6.	Sec.	5.7	b
7.	Sec.	5.1	a
8.	Sec.	5.14	F
9.	Sec.	5.1	T
10.	Sec.	5.3	F
11.	Sec.	5.1	F
12.	Sec.	5.10	F
13.	Sec.	5.1	F
14.	Sec.	5.1	T
15.	Sec.	5.1	low, wind, air
16.	Sec.	5.4	variable effects, lowering
17.	Sec.	5.12	poor, dense
18.	Sec.	5.1	volume, bleed, tensile
19.	Sec.	5.8	measure of elasticity, E
20.	Sec.	5.2	expansion, contraction, wetting, reversible

Chapter 6—Cracks and Blemishes

1.	Sec.	6.2	d
2.	Sec.	6.5	b
3.	Sec.	6.6	a
4.	Sec.	6.16	d
5.	Sec.	6.26	a
6.	Sec.	6.11	e
7.	Sec.	6.20	c
8.	Sec.	6.4	c
9.	Sec.	6.7	T
10.	Sec.	6.10	F
11.	Sec.	6.13	T
12.	Sec.	6.15	T
13.	Sec.	6.23	F
14.	Sec.	6.28	F
15.	Sec.	6.22	T

16.   Sec.   6.1   F
17.   Sec.   6.3   reinforcing, items embedded, aggregate particles, cracks
18.   Sec.   6.5   openings, reinforcing
19.   Sec.   6.9   previously placed, slabs, walls
20.   Sec.   6.21   diagnose, cause, extent
21.   Sec.   6.8   designed properly, sections, reinforcing
22.   Sec.   6.12   Bugholes, $1/2$ inch, dried paste, pressure
23.   Sec.   6.18   peeling, scaling
24.   Sec.   6.27   adhesives, mortar sand, one, adhesive, three, sand
25.   Sec.   6.19   spalling, 1 inch, 6 inches

Chapter 7—Portland Cement

1.   Sec.   7.8   c
2.   Sec.   7.4   c
3.   Sec.   7.2   d
4.   Sec.   7.4   c
5.   Sec.   7.8   d
6.   Sec.   7.2   T
7.   Sec.   7.10   T
8.   Sec.   7.9   F
9.   Sec.   7.8   F
10.   Sec.   7.10   T
11.   Sec.   7.5   F
12.   Sec.   7.4   sulfate-resistant, soil, ground, sulfate
13.   Sec.   7.8   hydrates, accelerates
14.   Sec.   7.5   IS, IS-A, S, IP, P
15.   Sec.   7.2   gypsum, setting time
16.   Sec.   7.6   iron, I, tinted, colored

Chapter 8—Aggregates

1.   Sec.   8.6   c
2.   Sec.   8.3   b
3.   Sec.   8.7   e
4.   Sec.   8.3   d
5.   Sec.   8.3   a
6.   Sec.   8.4   c
7.   Sec.   8.7   b
8.   Sec.   8.4   a
9.   Sec.   8.3   T
10.   Sec.   8.0   T
11.   Sec.   8.5   F
12.   Sec.   8.10   F
13.   Sec.   8.2   T
14.   Sec.   8.4   F
15.   Sec.   8.6   F

16.	Sec.	8.3	T
17.	Sec.	8.7	three, two, fines, dust
18.	Sec.	8.9	blast furnace
19.	Sec.	8.1	three, igneous, sedimentary, metamorphic
20.	Sec.	8.3	two tenths, two or three, one and one-half
21.	Sec.	8.3	three, one-half, one
22.	Sec.	8.4	heavy media, jigging, impact crusher, elastic fractionation
23.	Sec.	8.4	clay, silt, revolving, log washer, screw washer
24.	Sec.	8.6	few, high, cone, layers, closely, vertical
25.	Sec.	8.3	rough, cement paste, smooth

Chapter 9—Water and Admixtures

1.	Sec.	9.2	b
2.	Sec.	9.5	c
3.	Sec.	9.2	e
4.	Sec.	9.3	d
5.	Sec.	9.4	c
6.	Sec.	9.2	b
7.	Sec.	9.1	a
8.	Sec.	9.2	T
9.	Sec.	9.2	F
10.	Sec.	9.2	T
11.	Sec.	9.6	T
12.	Sec.	9.4	F
13.	Sec.	9.1	F
14.	Sec.	9.2	T
15.	Sec.	9.2	intermixed, manufacturers
16.	Sec.	9.4	volcanic tuff, volcanic ash, pumicite, obsidian
17.	Sec.	9.2	surfaces, natural, synthetic, polymers
18.	Sec.	9.2	absorption, capillary action
19.	Sec.	9.5	sulfate attack, alkali-silica, lowered heat
20.	Sec.	9.2	colorfast, chemically stable, setting time, strength

Chapter 10—Accessory Materials

1.	Sec.	10.1	a
2.	Sec.	10.3	e
3.	Sec.	10.6	a
4.	Sec.	10.1	e
5.	Sec.	10.3	c
6.	Sec.	10.4	F
7.	Sec.	10.5	F
8.	Sec.	10.7	T
9.	Sec.	10.2	embed, compressing
10.	Sec.	10.1	polyethylene, neoprene, butyl
11.	Sec.	10.3	resin, curing agent
12.	Sec.	10.2	sheet copper, rubbers, polyvinyl chloride

Chapter 11—Formwork

1.	Sec.	11.2	b
2.	Sec.	11.3	d
3.	Sec.	11.9	a
4.	Sec.	11.5	d
5.	Sec.	11.7	d
6.	Sec.	11.4	c
7.	Sec.	11.1	b
8.	Sec.	11.1	b
9.	Sec.	11.6	c
10.	Sec.	11.4	T
11.	Sec.	11.3	T
12.	Sec.	11.1	T
13.	Sec.	11.1	F
14.	Sec.	11.2	F
15.	Sec.	11.11	F
16.	Sec.	11.1	sagging, settlement, $1/4$ inch, span
17.	Sec.	11.1	joint, anchorages, 4 inches, lift
18.	Sec.	11.11	dirt, mortar, hardware, other material
19.	Sec.	11.3	shellac, lacquer, form, oil
20.	Sec.	11.9	locking devices, joined together, stacked

Chapter 12—Proportioning the Concrete Mixture

1.	Table	12-1	c
2.	Sec.	12.6	b
3.	Sec.	12.0	a
4.	Sec.	12.9	a
5.	Sec.	12.1	e
6.	Sec.	12.2	d
7.	Sec.	12.0	e
8.	Sec.	12.1	c
9.	Table	12.1	c
10.	Sec.	12.5	F
11.	Sec.	12.7	T
12.	Sec.	12.8	T
13.	Sec.	12.5	F
14.	Sec.	12.2	F
15.	Sec.	12.1	T
16.	Sec.	12.3	F
17.	Sec.	12.3	series, MSA, job
18.	Sec.	12.2	inside, between
19.	Sec.	12.3	depth, three-quarters, forms, one-fifth
20.	Sec.	12.3	all, two, seven, 14, 28
21.	Sec.	12.3	dry
22.	Sec.	12.5	water content, slump

Chapter 13—Testing and Controlling the Concrete

1.	Sec.	13.3	a
2.	Sec.	13.1	e
3.	Sec.	13.4	e
4.	Sec.	13.2	e
5.	Sec.	13.5	b
6.	Sec.	13.4	b
7.	Sec.	13.8	d
8.	Sec.	13.1	e
9.	Sec.	13.5	a
10.	Sec.	13.4	b
11.	Sec.	13.4	c
12.	Sec.	13.4	c
13.	Sec.	13.4	d
14.	Sec.	13.10	c
15.	Sec.	13.5	d
16.	Sec.	13.6	e
17.	Sec.	13.6	c
18.	Sec.	13.5	e
19.	Sec.	13.9	T
20.	Sec.	13.4	F
21.	Sec.	13.4	T
22.	Sec.	13.8	F
23.	Sec.	13.2	F
24.	Sec.	13.10	T
25.	Sec.	13.4	T
26.	Sec.	13.7	T
27.	Sec.	13.4	F
28.	Sec.	13.10	F
29.	Sec.	13.4	F
30.	Sec.	13.4	T
31.	Sec.	13.7	T
32.	Sec.	13.5	F
33.	Sec.	13.6	T
34.	Sec.	13.6	F
35.	Sec.	13.6	F
36.	Sec.	13.6	T
37.	Sec.	13.5	F
38.	Sec.	13.5	F
39.	Sec.	13.5	T
40.	Sec.	13.5	T
41.	Sec.	13.4	inches, low, high
42.	Sec.	13.4	leveled surface
43.	Sec.	13.4	top, bottom, 8", 12", 4"
44.	Sec.	13.2	observations, accuracy, reliability
45.	Sec.	13.1	voids, unit weight

46.   Sec.   13.4   slump, strength
47.   Sec.   13.5   two, 1 inch, three, three, four
48.   Sec.   13.3   verify, refute
49.   Sec.   13.2   representative
50.   Sec.   13.5   smooth, level, three, 25, $^5/_8$ inch, hemispherical

Chapter 14—Batching and Mixing the Concrete

1.   Sec.   14.1   b
2.   Sec.   14.3   b
3.   Sec.   14.9   c
4.   Sec.   14.3   b
5.   Sec.   14.10   c
6.   Sec.   14.7   d
7.   Sec.   14.8   d
8.   Sec.   14.1   a
9.   Sec.   14.6   e
10.   Sec.   14.9   b
11.   Sec.   14.1   F
12.   Sec.   14.1   T
13.   Sec.   14.10   T
14.   Sec.   14.9   F
15.   Sec.   14.1   T
16.   Sec.   14.8   T
17.   Sec.   14.9   F
18.   Sec.   14.9   F
19.   Table 14.1   F
20.   Sec.   14.9   T
21.   Sec.   14.9   batching, introducing, mixer drum
22.   Sec.   14.4   separately, cumulative, separate, scales
23.   Sec.   14.2   scale, indicator, signal, designed weight
24.   Sec.   14.5   interiors, gates, covered
25.   Sec.   14.9   aggregates, water, ice
26.   Sec.   14.2   semiautomatic, controls, automatic
27.   Sec.   14.6   blades, one, other, paths
28.   Sec.   14.1   truck, wrong, mud, clay
29.   Sec.   14.9   truck, contractor, ready-mix batch plant, concrete, loaded
30.   Sec.   14.8   bathtub, rounded contour
31.   Sec.   14.4   clean, dull, dirty, fulcrums

Chapter 15—Handling and Placing the Concrete

1.   Sec.   15.3   a
2.   Sec.   15.2   b
3.   Sec.   15.2   a
4.   Sec.   15.1   c
5.   Sec.   15.5   b

6.	Sec.	15.3	a
7.	Sec.	15.4	e
8.	Sec.	15.3	e
9.	Sec.	15.2	d
10.	Sec.	15.2	d
11.	Table	15.3	e
12.	Table	15.3	e
13.	Sec.	15.5	T
14.	Sec.	15.3	F
15.	Sec.	15.3	T
16.	Sec.	15.2	T
17.	Sec.	15.1	T
18.	Sec.	15.2	F
19.	Sec.	15.1	F
20.	Sec.	15.4	T
21.	Sec.	15.5	T
22.	Sec.	15.3	T
23.	Sec.	15.1	F
24.	Sec.	15.2	T
25.	Table	15.3	T
26.	Table	15.3	F
27.	Table	15.3	F
28.	Sec.	15.1	regular, smooth
29.	Sec.	15.2	segregation, concrete, consistency
30.	Sec.	15.2	obstructions
31.	Sec.	15.3	high, mix, dry, sun, pumping
32.	Sec.	15.5	forms, reinforcing
33.	Sec.	15.1	anchor bolts, pipes, conduits, catch basins
34.	Sec.	15.3	P150, 100, 3000
35.	Sec.	15.5	spaced, consolidation, 5 to 15

Chapter 16—Slabs on Ground

1.	Sec.	16.3	a
2.	Sec.	16.3	c
3.	Sec.	16.1	e
4.	Table	16.1	d
5.	Sec.	16.2	c
6.	Sec.	16.1	a
7.	Sec.	16.1	b
8.	Sec.	16.2	b
9.	Sec.	16.1	d
10.	Sec.	16.2	b
11.	Sec.	16.2	T
12.	Sec.	16.1	F
13.	Sec.	16.1	F
14.	Sec.	16.3	F

15.	Sec.	16.2	F
16.	Sec.	16.1	F
17.	Sec.	16.2	T
18.	Sec.	16.2	T
19.	Sec.	16.2	F
20.	Sec.	16.4	subgrade, building
21.	Sec.	16.1	6
22.	Sec.	16.1	one day, damp
23.	Sec.	16.1	$^1/_8$, foot, $^1/_4$, foot, ponding
24.	Sec.	16.2	bulkhead, construction, predetermined
25.	Sec.	16.3	curing, slump, bleed water

Chapter 17—Finishing and Curing the Concrete

1.	Sec.	17.8	b
2.	Sec.	17.4	a
3.	Sec.	17.3	d
4.	Sec.	17.8	c
5.	Sec.	17.1	a
6.	Sec.	17.3	d
7.	Sec.	17.4	d
8.	Sec.	17.8	a
9.	Sec.	17.6	d
10.	Sec.	17.5	b
11.	Sec.	17.5	e
12.	Sec.	17.1	d
13.	Sec.	17.2	T
14.	Sec.	17.3	T
15.	Sec.	17.1	T
16.	Sec.	17.8	F
17.	Sec.	17.6	F
18.	Sec.	17.7	F
19.	Sec.	17.4	F
20.	Sec.	17.4	T
21.	Sec.	17.4	F
22.	Sec.	17.1	F
23.	Sec.	17.3	T
24.	Sec.	17.6	F
25.	Sec.	17.4	T
26.	Sec.	17.1	edging, chipping, damage
27.	Sec.	17.4	white, pigment, silica sand
28.	Sec.	17.1	high-carbon, 10, 20, 3, 5
29.	Sec.	17.8	burlap, cotton mats, fabric
30.	Sec.	17.3	abrasion, impact, after, disappeared
31.	Sec.	17.6	water, liquid membrane-forming, sheet materials, blankets
32.	Sec.	17.1	steel, bronze, malleable iron, 6, upturned
33.	Sec.	17.3	hard, tough, quartz, granite

34.   Sec.   17.5   spraying, nontoxic, damp, harmfully
35.   Sec.   17.8   strength, other propertied, first few hours

Chapter 18—The Steel Reinforcement

1.   Sec.   18.5   b
2.   Sec.   18.4   e
3.   Sec.   18.7   d
4.   Sec.   18.2   a
5.   Sec.   18.2   d
6.   Sec.   18.5   e
7.   Sec.   18.2   c
8.   Sec.   18.5   a
9.   Sec.   18.2   c
10.   Sec.   18.2   b
11.   Sec.   18.2   d
12.   Sec.   18.4   a
13.   Sec.   18.2   b
14.   Sec.   18.2   d
15.   Table   18.2   c
16.   Figure   18-4   c
17.   Figure   18-17B   d
18.   Figure   18-17B   a
19.   Figure   18-17B (*Concrete Manual* Figure 18-36)   d
20.   Sec.   18.5   c
21.   Sec.   18.8   d
22.   Sec.   18.5   c
23.   Sec.   18.5   e
24.   Table   18.6   c
25.   Table   18.6   c
26.   Table   18.6   c
27.   Table   18.7   d
28.   Sec.   18.2   T
29.   Sec.   18.3   F
30.   Sec.   18.5   T
31.   Sec.   18.5   T
32.   Sec.   18.2   T
33.   Sec.   18.4   F
34.   Sec.   18.2   F
35.   Sec.   18.2   T
36.   Sec.   18.7   T
37.   Sec.   18.3   T
38.   Sec.   18.5   F
39.   Sec.   18.4   F
40.   Sec.   18.2   F
41.   Sec.   18.2   T
42.   Sec.   18.2   F

43.	Table	18.2	T
44.	Figure	18-4	F
45.	Sec.	18.2	T
46.	Sec.	18.4	T
47.	Sec.	18.5	F
48.	Sec.	18.5	F
49.	Table	18.7	T
50.	Sec.	18.9	F
51.	Sec.	18.9	T
52.	Sec.	18.5	prohibited, engineer, building official
53.	Sec.	18.2	one, 60, two, 75
54.	Sec.	18.5	beams, midspan, cover
55.	Sec.	18.7	freezing, thawing, de-icing salts
56.	Sec.	18.4	platforms, supports, damage, dirt, mud, rust
57.	Sec.	18.3	reinforcing, size, length, straight, bent
58.	Sec.	18.3	measurements, tolerances, one
59.	Sec.	18.5	chairs, ties, hangers, supports
60.	Sec.	18.4	building official, engineer, slowly, slowly
61.	Sec.	18.1	expansion, contraction, changes, cracking
62.	Sec.	18.2	designer, plans, bar lists

Chapter 19—Hot and Cold Weather Concreting

1.	Sec.	19.1	c
2.	Sec.	19.3	c
3.	Sec.	19.4	b
4.	Sec.	19.4	b
5.	Sec.	19.2	a
6.	Figure	19-3	b
7.	Sec.	19.3	e
8.	Sec.	19.1	T
9.	Sec.	19.4	F
10.	Sec.	19.2	T
11.	Sec.	19.1	T
12.	Sec.	19.2	F
13.	Figure	19-3	T
14.	Sec.	19.4	should not, mixing water
15.	Sec.	19.3	40 and 70, standard
16.	Sec.	19.2	uniformity, quality, overmixing
17.	Sec.	19.1	negative, reduced, aggressive
18.	Sec.	19.3	protection, hardened, little strength

Chapter 20—Precast and Prestressed Concrete

1.	Sec.	20.10	b
2.	ACI 318 Sec. 18.16.2		b
3.	Sec.	20.10	c

4.	Sec.	20.6	c
5.	Sec.	20.10	c
6.	ACI 318 Sec. 18.20.1		e
7.	ACI 318 Sec. 18.16.2		e
8.	ACI 318 Sec. 18.20.1		d
9.	ACI 318 Sec. 18.16.1		c
10.	Sec.	20.1	a
11.	Sec.	20.5	c
12.	Sec.	20.7	b
13.	Sec.	20.4	b
14.	Sec.	20.10	a
15.	Figure	20-29	d
16.	Sec.	20.10	a
17.	Sec.	20.6	T
18.	Sec.	20.0	T
19.	Sec.	20.6	F
20.	Sec.	20.6	T
21.	Sec.	20.1	F
22.	Sec.	20.6	F
23.	Sec.	20.10	T
24.	Sec.	20.4	T
25.	Sec.	20.6	F
26.	Sec.	20.7	F
27.	Sec.	20.3	F
28.	Sec.	20.6	T
29.	Sec.	20.6	F
30.	Sec.	20.10	T
31.	Sec.	20.3	T
32.	Sec.	20.6	F
33.	Sec.	20.6	F
34.	Sec.	20.6	T
35.	Sec.	20.10	T
36.	Figure	20-29	F
37.	Figure	20-29	T
38.	Table	20.1	high, low, smooth, shaped
39.	Sec.	20.10	before
40.	Sec.	20.2	fabricator, manufacturer
41.	Sec.	20.8	identical
42.	Sec.	20.6	anchors, tendons, ends
43.	Sec.	20.6	prior, concrete, after, concrete
44.	Sec.	20.6	stress relieved, loss, yield
45.	Sec.	20.7	embedded, made, anchorage, attachment
46.	Sec.	20.10	corrosion-preventive, lubricant, sheathing
47.	Sec	20.6.	28,000,000; eight

Chapter 21—Lightweight and Heavyweight Concrete

1.	Sec.	21.4	b
2.	Sec.	21.2	b
3.	Sec.	21.2	e
4.	Sec.	21.1	c
5.	Sec.	21.3	d
6.	Sec.	21.2	F
7.	Sec.	21.2	F
8.	Sec.	21.4	F
9.	Table	21.1	F
10.	Sec.	21.2	T
11.	Sec.	21.2	crushes, screens, small, coal, coke
12.	Sec.	21.3	air, gas, remaining
13.	Sec.	21.2	vibration, internal vibrators, segregation
14.	Sec.	21.2	absorbed, specific gravity, moisture
15.	Sec.	21.3	foaming agent, water, paddle, pan

Chapter 22—Special Concreting Techniques

1.	Sec.	22.1	d
2.	Sec.	22.2	a
3.	Sec.	22.7	d
4.	Sec.	22.13	b
5.	Sec.	22.13	b
6.	Sec.	22.4	a
7.	Sec.	22.14	c
8.	Sec.	22.8	e
9.	Sec.	22.1	a
10.	Sec.	22.5	d
11.	Sec.	22.13	b
12.	Sec.	22.2	e
13.	Sec.	22	c
14.	Sec.	22.13	a
15.	Sec.	22.7	d
16.	Sec.	22.13	c
17.	Sec.	22.10	c
18.	Sec.	22.13	b
19.	Sec.	22.1	e
20.	Sec.	22.13	b
21.	Sec.	22.15	c
22.	Sec.	22.16	d
23.	Sec.	22.16	d
24.	Sec.	22.17	a
25.	Sec.	22.16	d
26.	Sec.	22.18	d
27.	Sec.	22.19	e

28.	Sec.	22.19	e
29.	Sec.	22.3	T
30.	Sec.	22.7	F
31.	Sec.	22.13	T
32.	Sec.	22.11	F
33.	Sec.	22.7	F
34.	Sec.	22.12	F
35.	Sec.	22.13	T
36.	Sec.	22.6	T
37.	Sec.	22.13	F
38.	Sec.	22.13	T
39.	Sec.	22.2	T
40.	Sec.	22.13	F
41.	Sec.	22.2	T
42.	Sec.	22.4	T
43.	Sec.	22.1	F
44.	Sec.	22.15	F
45.	Sec.	22.15	F
46.	Sec.	22.16	T
47.	Sec.	22.15	F
48.	Sec.	22.17	F
49.	Sec.	22.15	F
50.	Sec.	22.16	T
51.	Sec.	22.18	F
52.	Sec.	22.18	T
53.	Sec.	22.18	T
54.	Sec.	22.11	T
55.	Sec.	22.7	mixing water, chamber, compressed air
56.	Sec.	22.11	gun, predetermined lengths, sand, cement
57.	Sec.	22.13	one, age
58.	Sec.	22.3	tremies, buckets, pumping
59.	Sec.	22.8	dry-pack mortar, grout
60.	Sec.	22.13	oxide, test panels
61.	Sec.	22.7	removed, sandblasted, doweled, bolted
62.	Sec.	22.13	hydrated cement, limestone, marble
63.	Sec.	22.10	three to four, freezing, thawing, sulfate, abrasion, decreased
64.	Sec.	22.4	slowly, previously placed

Chapter 23—Waterproofing and Dampproofing

1.	Sec.	23.2	b
2.	Sec.	23.3	a
3.	Sec.	23.5	c
4.	Sec.	23.1	c
5.	Sec.	23.4	d

6.   Sec.   23.4   F
7.   Sec.   23.3   F
8.   Sec.   23.2   T
9.   Sec.   23.3   T
10.   Sec.   23.3   T
11.   Sec.   23.3   clean, dry, smooth, filled, mortar, removed
12.   Sec.   23.5   unsound, encrustation, V-grooved, top, lowest
13.   Sec.   23.2   one, more, hot, cold, glass, plastic
14.   Sec.   23.3   grade, walls, floors passage of liquid
15.   Sec.   23.1   well-graded, low porosity, rounded

Chapter 24—Introduction to Inspection

1.   Sec.   24.1   c
2.   Sec.   24.3   c
3.   Sec.   24.1   a
4.   Sec.   24.8   e
5.   Sec.   24.4   d
6.   Sec.   24.12   a
7.   Sec.   24.10   c
8.   Sec.   24.11   e
9.   Sec.   24.6   b
10.   Sec.   24.2   b
11.   Sec.   24.2   d
12.   Sec.   24.3   b
13.   Sec.   24.1   T
14.   Sec.   24.9   T
15.   Sec.   24.1   F
16.   Sec.   24.5   T
17.   Sec.   24.12   T
18.   Sec.   24.7   F
19.   Sec.   24.7   T
20.   Sec.   24.12   F
21.   Sec.   24.7   F
22.   Sec.   24.2   T
23.   Sec.   24.6   F
24.   Sec.   24.2   F
25.   Sec.   24.10   dry, oven, hot plate
26.   Sec.   24.9   quality control, specifications, properties
27.   Sec.   24.3   ASTM, 4, 7
28.   Sec.   24.7   service, mill, factory, tests, independent
29.   Sec.   24.5   job, construction documents
30.   Sec.   24.1   suggestions, instructions, workers
31.   Sec.   24.10   trash, mud, contaminates

Chapter 25—Inspection of Concrete Construction

1.	Sec.	25.3	c
2.	Sec.	25.9	b
3.	Sec.	25.10	a
4.	Sec.	25.8	b
5.	Sec.	25.2	c
6.	Sec.	25.9	d
7.	Sec.	25.1	b
8.	Sec.	25.5	d
9.	Sec.	25.9	e
10.	Sec.	25.9	c
11.	Sec.	25.4	F
12.	Sec.	25.6	T
13.	Sec.	25.0	F
14.	Table	25.1	T
15.	Sec.	25.8	F
16.	Sec.	25.10	T
17.	Sec.	25.5	T
18.	Sec.	25.2	T
19.	Sec.	25.6	F
20.	Sec.	25.9	T
21.	Sec.	25.9	F
22.	Sec.	25.9	T
23.	Sec.	25.9	T
24.	Sec.	25.10	F
25.	Sec.	25.10	F
26.	Sec.	25.1	batching, mixing, placing, finishing
27.	Sec.	25.7	permit, size
28.	Sec.	25.9	inspection reports, design drawings, specifications
29.	Sec.	25.9	amendments, procedures, requirements
30.	Table	25.1	calibration, sequence
31.	Table	25.1	proper length, slope, length
32.	Sec.	25.4	tamping, rolling, aggregate
33.	Sec.	25.2	load ticket, thoroughly mixed, time limits

Chapter 26—Quality Control

1.	Sec.	26.1	b
2.	Sec.	26.2	d
3.	Sec.	26.2	a
4.	Table	26.1	a
5.	Sec.	26.1	c
6.	Sec.	26.3	a
7.	Sec.	26.3	c
8.	Sec.	26.1	T
9.	Sec.	26.0	T

10.	Sec.	26.1	T
11.	Sec.	26.1	F
12.	Sec.	26.2	F
13.	Sec.	26.1	concrete strength, analysis, sand, continuing
14.	Sec.	26.2	quality, strength, results
15.	Sec.	26.2	percentage, lower, design
16.	Sec.	26.1	does not, savings
17.	Sec.	26.2	uniformity, quality, materials, manufacture, controlled